FAMILY MATH

Jean Kerr Stenmark
Virginia Thompson
and
Ruth Cossey

Illustrated by Marilyn Hill

The Lawrence Hall of Science is a public science center, teacher inservice institution, and research unit in science education at the University of California, Berkeley. For many years, it has developed curricula and teaching strategies to improve mathematics and science education at all levels, and to increase public understanding of, and interest in, science and mathematics.

For information about the FAMILY MATH program or additional copies of the book, contact:

**Lawrence Hall of Science
University of California
Berkeley, CA 94720
Attn: FAMILY MATH**

**(510) 642-1823 program
(510) 642-1910 books
(510) 643-5757 FAX**

Credits:

Design & Illustration: MARILYN HILL

Design Assistant: CONNIE TORII

Consultant: CAROL LANGBORT

Typographer: ANN FLANAGAN TYPOGRAPHY—Linda Davis

Printer: U.C. PRINTING SERVICES

Career Story Cards: WENDY WARREN

The material was prepared with the support of the U.S. Department of Education (Fund for the Improvement of Postsecondary Education) and the Carnegie Corporation of New York. However, any opinions, findings, conclusions, or recommendations expressed herein are those of the authors and do not necessarily reflect the views of the U.S. Department of Education or the Carnegie Corporation.

ISBN 0-912511-06-0

CONTENTS

ACKNOWLEDGMENTS

The authors of this book are pleased to acknowledge help and support from many sources.

The Fund for the Improvement of Postsecondary Education of the U.S. Department of Education provided funds for the three-year grant that enabled us to develop, test, and disseminate FAMILY MATH. Without this, there would not have been a program.

We are grateful to the Carnegie Corporation of New York for funding the publication and dissemination of this book which we hope will encourage people throughout this country and abroad to become part of FAMILY MATH.

The FAMILY MATH course introduces parents and their children to good ideas that help them improve their mathematics skills and gain appreciation for mathematics. The original instructors of the FAMILY MATH classes planned carefully, using available materials, modifying where necessary, and creating new activities as needed. Many mathematics educators were extremely generous in allowing us to use their activities for the course.

A number of the activities included in the book are our variations of those used by colleagues or modifications of ideas that have appeared in various mathematics education publications. Many are considered classics.

We have been fortunate to receive inspiration, suggestions, ideas, and activities from many individuals, including Marilyn Burns, Regina Comaich, Ernestine Camp, Jill DeJean, Wallace Judd, Carol Langbort, Bill Ruano, Hal Saunders, Dale Seymour, Kathryn Standifer, Paula Symonds, Tamas Varga, Marion Walter, and the late Robert Wirtz.

The following publications and projects have been rich sources of ideas for our program and would be of value for others who want more activities. *The Arithmetic Teacher, Games Magazine, The Lane County Mathematics Project, The Oregon Mathematics Teacher, The Mathematical Reasoning Improvement Study, The Miller Math Project, and Seedbed.* Please see the Resource List on page 311 for additional publications and publisher addresses.

We deeply appreciate the continued ideas, support, and counsel of present and past colleagues from the EQUALS Project and the Lawrence Hall of Science at the University of California, Berkeley:

Lynne Alper, Susan Arnold, Diane Downie, Tim Erickson, Doug FitzGerald, Sherry Fraser, Kay Gilliland, Ellen Humm, Helen Joseph, Alice Kaseberg, Nancy Kreinberg, Helen Raymond, Diane Resek, Twila Slesnick, and Elizabeth Stage.

And most of all, we thank the teachers, parents, and children who have been part of the exciting development of FAMILY MATH.

Jean Kerr Stenmark
Virginia Thompson
and
Ruth Cossey

Berkeley, California
1986

INTRODUCTION

FAMILY MATH: WHAT IS IT?

For many years, the EQUALS program at the Lawrence Hall of Science at the University of California, Berkeley, has worked with teachers who wanted to improve mathematics teaching and learning in their classrooms and schools, providing activities and methods to help students succeed in mathematics, especially females and all students of color.

The EQUALS program has three strands—Awareness, Confidence, and Encouragement. Teachers and students are made aware of the need for mathematics and the options for young people; confidence is built by providing strategies for success in mathematics; encouragement involves motivating students to continue studying mathematics and to consider a wide variety of careers.

Many of the teachers who came to the program asked us to give them ideas and materials for parents to use at home to help their children in mathematics. They told us that parents were frustrated in not knowing enough about their children's math program to help them or in not understanding the mathematics their children were studying.

We thought we could do something about this by creating a separate program—FAMILY MATH—that would focus entirely on parents and children learning mathematics together. A grant from the Fund for the Improvement of Postsecondary Education (U.S. Department of Education) gave us the time and money to develop our ideas and test them out in a number of communities, with families from inner cities, suburbs, and rural areas. This book is the result of many years' work in providing FAMILY MATH courses.

To find out more about the history and impact of FAMILY MATH, and how you can get involved, call or write for our FAMILY MATH/ MATEMATICA PARA LA FAMILIA booklet and program information.

THE FAMILY MATH COURSE

A typical FAMILY MATH course includes six or eight sessions of an hour or two and gives parents and children (Kindergarten through grade 8) opportunities to develop problem-solving skills and to build an understanding of mathematics with "hands-on" materials.

By **problem-solving skills** we mean ways in which people learn how to think about a problem using such strategies as looking for patterns, drawing a picture, working backwards, working with a partner, or eliminating possibilities. Having a supply of strategies allows a choice of ways to start looking at a problem, relieving the frustration of not knowing how or where to begin. The more strategies you have, the more confident you become, the more willing you are to tackle new problems, and the better problem solver you become.

By **"hands-on" materials,** we mean concrete objects—like blocks, beans, pennies, toothpicks—that are used to help children understand what numbers and space mean, and to help all of us solve problems. Traditionally, these materials are used mostly in the early elementary years and paper-and-pencil mathematics becomes the rule after second or third grade. This is unfortunate, since much of mathematics can best be explained and understood using the **tools of manipulative materials and models;** and, in fact, many research and applied mathematicians do just that.

Parents in FAMILY MATH classes are also given overviews of the mathematics topics at their children's grade levels and explanations of how these topics relate to each other. Sometimes, the curriculum begins to make sense simply by knowing how one concept builds the base for the next.

To ensure that the **reason** for studying mathematics is clear, men and women working in math-based occupations come to FAMILY MATH classes to talk about how math is used in their jobs and those of the people with whom they work. These role models are often extremely important for adolescents who are beginning to think they will never use math as the rock star or professional athlete they expect to become. Career activities used in FAMILY MATH classes highlight the twists and turns that most of our lives take and how far we are from what we thought at age 14 we would become.

WHAT'S IN A FAMILY MATH COURSE?

Courses are usually taught by grouped grade levels such as K-3, 4-6, or 6-8; although many variations occur depending upon the teacher and the families. Materials for each course are based on the school mathematics program for those grade levels and reinforce the concepts that are introduced throughout the curriculum.

Topics included in FAMILY MATH classes fall into the general categories of arithmetic, geometry, probability and statistics, measurement, estimation, calculators, computers, logical thinking, and careers.

Some of these subjects may seem unfamiliar to you, and you may wonder how they fit into the standard school curriculum. In fact, they are in most textbooks, but limited to a page or two, and often relegated to the back of the book and ignored. However, a look at mathematics in daily life will reveal that arithmetic is by

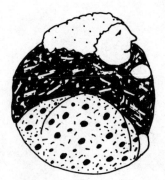

no means the only concept used; the other topics share equal importance.

As students progress through mathematics, it is essential that they develop an ability to visualize spatial relationships (geometry), to approximate (estimation), to interpret data (probability and statistics), and to reason mathematically (logical thinking). Time spent on topics other than arithmetic need not detract from students' learning of "basics." In fact, the opposite may be true—a student who begins to understand probability, for example, may find addition and subtraction suddenly easier. Making graphs can lead to an understanding of ratios and other relative values. Estimation requires a kind of thinking that leads to making sense of numbers. These topics are inherently interesting and often can attract a student who has had no interest or success in mathematics previously. And, should we add, they are fun to teach.!

WHO TEACHES FAMILY MATH?

Anyone who is enthusiastic, kind, and not afraid. Thousands of families have taken a FAMILY MATH course taught by their child's classroom teacher, a parent, a teacher aide, a community college instructor, or a retired person. Naturally, it may be easier for an experienced teacher to put a class together, but parents have found it to be a wonderful way to become involved in the community and their child's school. The information to help you organize a FAMILY MATH class is found in Appendix A.

WHAT'S IN THIS BOOK?

There are enough activities in this book to teach a FAMILY MATH class, with information in Appendix A to help organize the class. The book has been written, however, with the parent at home in mind. We have tried to make the presentation of the activities and the directions clear enough so that a parent can use the book without attending a class. This will make the materials easy to use in a class, since the parents who have a chance to come to a class will expect to continue activities at home and will need to refer back to the directions.

We recommend that parents take advantage of any opportunity to work with teachers and/or other parents to do the activities together. We believe that it is extremely important for children to have opportunities to talk about mathematics with others, and equally important for parents to be able to talk with other adults about mathematics.

Dear FAMILY MATH

My husband and I are interested in fostering an interest in mathematics in our daughter and any ideas or lists of reference materials which you might have would be most welcome. We are beginning to think about this very early (she is 8 months old) but we know that her early years will be very important and that we must be prepared.

Not many of you will be quite this eager to begin doing mathematics with your baby, but FAMILY MATH is a way for parents and children to spend time together doing something that's fun, challenging, and important.

FAMILY MATH helped me remember some of what I forgot, but mostly it helped me to help my children without getting upset with them.

It's also a way for parents and children to talk to and help each other while they develop an understanding of mathematical concepts and strategies.

I want to give my children every possible chance to be able to feel comfortable with math, which hasn't been the case for me. I want to be able to be a guiding force and to demonstrate to them how easy math can be.

FAMILY MATH involves parents and children in problem solving, experimenting, and discovering together. But, most importantly, as one parent told us:

It's better to use it, see it, hear it, than read about it.

AND NOW THE FAMILY MATH BOOK

Now you are ready to try the activities in this book. Look for those that are at appropriate grade levels for your family, on topics that you are interested in or that your children are working on at school. DO NOT feel that you have to begin at the beginning and work your way through from back to front. Try doing the first activity from each chapter, or one from the middle of each chapter. Then go back and try several more from your favorite group. Or open the book at random and try whatever you find.

Above all, have fun!!

Sample Activity Page

Why

This gives a rationale for doing the activity, or its connection to the curriculum

How

☐ Gives the description of the activity and directions for preparing and doing it.

☐ Each step is marked with a ☐.

☐ There is also space in the margins or at the bottom of the pages to write your own notes about how your family does the activities. The book is meant to be used, so please write in it.

☐ There are many game boards and some worksheets which you may want to have duplicated so you won't have to tear out the pages. Gameboards and markers can be glued to cardboard for durability and ease of handling, or copied onto heavy paper.

► *Information in italics with lines above and below represents something like the voice of a teacher speaking, telling more about the mathematics involved in the activity, or giving advice about what to expect from your children.* ◄

More Ideas

☐ Some additional suggestions to extend or adapt the activity.

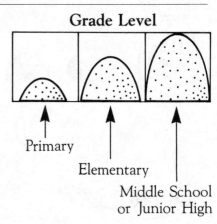

Grade Level

Primary

Elementary

Middle School or Junior High

TOOL KIT

A list of materials needed for this activity (Also see TOOL KIT following each chapter introduction.)

Chapter I

A MATHEMATICAL ENVIRONMENT

A MATHEMATICAL ENVIRONMENT

As parents, teachers, aunts, uncles, grandparents, or friends, most of us know that if we take a child to the library once a week and read aloud to her or him a lot, it provides a good start toward enjoyment of reading and therefore toward reading well.

But what do we know about helping children enjoy mathematics? Is it learning the times tables? Completing pages of long division problems? Or is there a mathematics library somewhere that has motivating mathematical books to be read?

What do you remember of your own mathematical education? Was it a pleasure or a pain? What topics did you study? Was it all addition, subtraction, multiplication, division and percents? How do you feel now when your children ask you for help with math homework?

This book is about helping all kinds of people love mathematics— parents, children, teachers, neighbors—everybody! Mathematics is more than arithmetic—it is beautiful and fascinating and exciting and meant to be enjoyed. Children (and grownups) who have explored geometry, probability, statistics, measurement, and logic, and who have learned to estimate and to see patterns and recognize relationships will be able to regard difficult problems as a challenge rather than as drudgery. What a gift for your family!

FAMILY MATH ACTIVITIES

Whether they are done at home or in a class, FAMILY MATH activities are meant to be fun. There is no rush or need for immediate mastery of ideas. There is no test at the end, and nobody asks for a checklist of skills learned. You can take your time, continue an activity as long as children are interested, try new things, learn new concepts along with the children.

Especially at home, this is a wonderful opportunity to create an environment that makes math seem pretty special and attractive. This chapter includes many suggestions to help bring this about.

DOING MATHEMATICS AT HOME

Here are some ideas for you to consider as you and your family are doing mathematics at home:

Let your children know that you believe they can succeed.

Let them see you enjoying the activities, liking mathematics. Children tend to emulate their parents, and if a parent says "You know, this is really interesting!" that becomes the child's model.

Be ready to talk with your children about mathematics and to listen to what they are saying. Even when you yourself don't know how to solve a problem, asking a child to explain the meaning of each part of the problem will probably be enough to find a strategy.

Be more concerned with the *processes* of doing mathematics than with getting a correct answer. The answer to any particular problem has very little importance, but knowing **how** to find the answer is a lifetime skill.

Try not to tell children *how* to solve the problem. Once they have been told how to do it, thinking usually stops. Better to ask them questions about the problem and help them find their own methods of working it through.

Practice estimation with your children whenever possible. Estimation helps the thinking about a problem that precedes the doing, and is one of the most useful and "sense-making" tools available.

Provide a special place for study, allowing your child to help you gear the study environment to his or her learning style. Some kids really do work better sprawled on the floor or bed, or with a musical background. There are no hard and fast rules.

Encourage group study. Open your home to informal study groups. Promote outside formal study groups related perhaps to scouts, church, or school organizations. This will be especially important as your children grow older.

Expect that homework will be done. Look at the completed work regularly. But **try to keep your comments positive.** Don't become a drill sergeant. Praise your child for asking questions, and look for places where you can ask questions about the work. To be successful, your child will need to study 30 to 60 hours a week in college, at least an hour and a half each day in middle school, and probably 20 minutes a day in elementary grades. The experts tell us that there is a high correlation between success in mathematics and the amount of homework done.

Don't expect that all homework will be easy for your child or be disappointed that it seems difficult. Never indicate that you feel your child is stupid. This may sound silly, but sometimes loving, caring parents unintentionally give their kids the most negative messages: for example, "Even your little sister Stephanie can do that," or "Hurry up, can't you see that the answer is ten?" or

"Don't worry, math was hard for me, too—and besides, you'll never use it!" or "How come you got a B in math when you could get A's in everything else?"

Seek out positive ways to support your child's teacher and school. Join the parent group. Offer to help find materials or role models. Accompany field trips. Avoid making negative comments about the teacher or the school in front of your child—your child needs to maintain a good feeling about the school.

Ask the teacher to give you a course outline or a list of the expectations for each class or subject of study. They should be available at the beginning of the school year or at a Back to School Night. These will help you know how your child is doing. There is a sample list in Appendix B.

Find time to sit in on your child's classes. Make an appointment with the school and attend each scheduled class for a day. Go to school events such as Open House and Back to School Night.

Look carefully at the standardized test results, and ask about any test scores that may indicate a skill deficiency or a special talent. But do not use these test scores as your **primary** means of assessment. The teacher's observations and your own will be much more valuable. Some of the most important attributes, such as sticking with a problem, or having many effective strategies to use, are not tested with paper and pencil.

Ask the teacher at the beginning of the term how place-ment decisions are made for subsequent courses, especially if your child is in the upper grades or is changing schools.

Try not to drill your child on math content or create hostilities by insisting that math work be done at any one specific time or in a specific way. Don't use math work as a punishment. Parents and adolescents have enough things that may create friction without adding math to the list.

Model persistence and pleasure with mathematics. Include enrichment, recreational mathematics in your family routine. Try to introduce math ideas (with a light touch!) at the dinner table, or while traveling, even to the grocery store.

Above all, enjoy mathematics!

Chapter II

BEGINNING

TOOL KIT

Paper

Pencils and pens

Scissors

Counting objects and game markers such as pennies, beans, bottle caps, small blocks, buttons

Egg cartons

Crayons

Dice

**Graph paper
(see page 79-82)**

Toothpicks

Straws

3″ × 5″ cards

A metric ruler

A calculator

BEGINNING

The activities in this chapter are known as "Openers." They are good activities to present at the beginning of a FAMILY MATH class or to use when a family begins to explore mathematics.

The first experiences in FAMILY MATH should be highly interesting, with games or problems that involve talking about what you are doing, using manipulative materials that can be moved around to try out different arrangements, if possible. Usually, it is better to minimize explaining at the start, moving quickly into the activity. More explanations and comments may follow at the end of the activity.

The first three activities are particularly suited for young children. They provide practice and reinforcement of arithmetic skills and number sense. Children will learn to understand the meaning of **five,** for example, in concrete terms rather than as an abstract numeral or a word in a sequence.

The other activities are appropriate for almost any age, and include topics other than arithmetic, such as grid games, logic, and geometric shapes. Some involve estimation and mental arithmetic, giving children and adults the feeling of power that comes with thinking about what the numbers really are, instead of performing rote paper and pencil drills. The initial activities will use concrete materials to help understand the meaning of numbers.

On the Dot

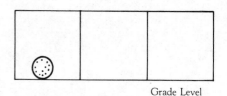

Why

To understand numbers

How

☐ Make four large dots on ten or fifteen pieces of paper. Set out some small objects.

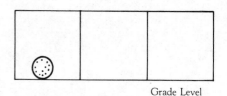

Grade Level

TOOLS

Paper
Pens
Counting objects

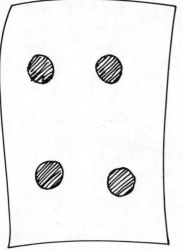

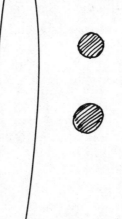

☐ Ask your child to count out four objects (pennies, beans, etc.) for each sheet, placing the objects on the dots while saying "1, 2, 3, 4."

☐ Practice with these sheets for several days, then ask your child to place four objects on ten or fifteen blank sheets. When your child can do this accurately, go on to five objects.

☐ This repetition and concentration on one number at a time is important in building a child's inner sense of the reality of each number. Don't rush to go on to the next number.

► *Most young children will be able to count by rote to 10 or 20, but may not be able to give you 10 or 20 pennies with accuracy.* ◄

Egg Carton Numbers

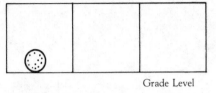

Grade Level

TOOLS

Egg carton
78 beans

Why

To gain experience with numbers

How

☐ Label the sections of an egg carton with the numbers 1 through 12.

☐ Give your child 78 beans (or other small objects) and ask him/her to count them into the sections of the carton, according to the numbers. There should be one bean in the section marked "1," two beans in the section marked "2," and so on.

☐ If the counting is accurate, the child will use exactly 78 beans. Can you explain why?

Odd or Even

Why

To understand numbers

How

- ☐ Have your child take a handful of beans or other small objects.
- ☐ Count them with your child.
- ☐ Then help your child arrange them in pairs to find out if the number is odd or even. An odd number will have one bean left over; an even number will come out even, with no beans left over.
- ☐ Keep a record of what happens. Do you see a pattern?

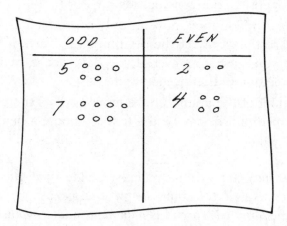

- ☐ To picture the pattern, write the numbers in a row. Color the odd numbers blue and the even numbers red.

► *Oddness and evenness are important concepts for understanding fractions, algebra and other advanced mathematics.* ◄

TOOLS

Record sheet
Beans or small objects
Blue and red crayons

Grade Level

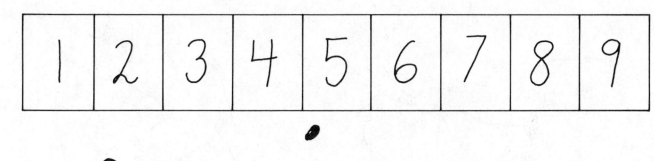

Animal Crossing

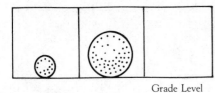

Grade Level

TOOLS

Animal Crossing board
   I or II

4 animal markers

1 die marked
   1, 1, 2, 2, 3, 3,

A game for
2–4 players

Why

To develop an understanding of two-dimensional representations by moving game pieces on a grid

How

☐ Each person chooses a different side of the board, and places his or her marker anywhere on the edge of that side.

☐ Take turns rolling the die. You may move the number of squares indicated by the dice, or you may move fewer squares.

☐ On your first move, you may move into any square on your side of the board, continuing from there. Your goal is to reach the side opposite your starting side.

☐ On each turn, you may move in only one direction. You may change directions only at the beginning of a turn. If you reach a barrier, you must stop, even if you have not moved the number of squares the dice showed.

☐ Play until all of the animals have moved across the board. The last person to finish may be the first to choose a new side for the next game.

► *This game may be played many times. Children will gain counting experiences, as well as learning simple strategies as they plan each move. Familiarity with a grid is important in later mathematics, especially geometry and calculus. Children may make up their own boards to extend the activity.* ◄

ANIMAL CROSSING I

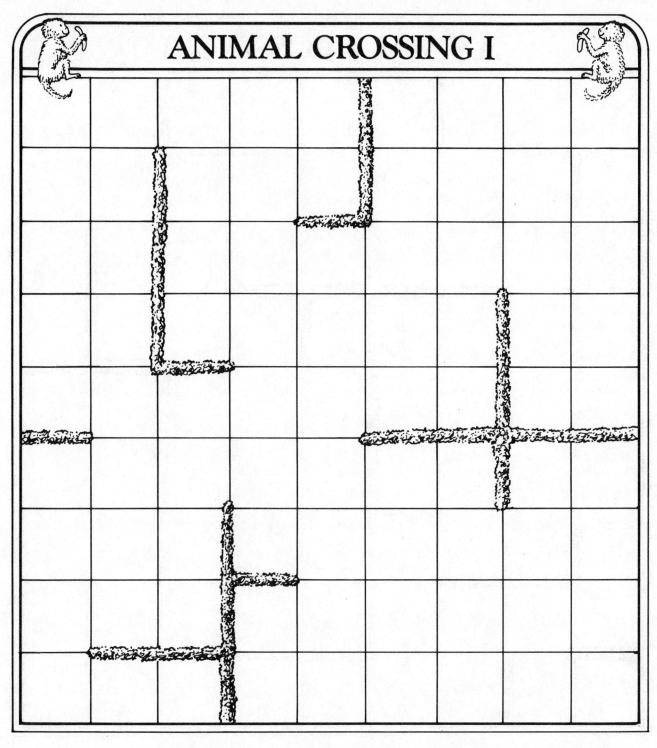

Markers

ANIMAL CROSSING II

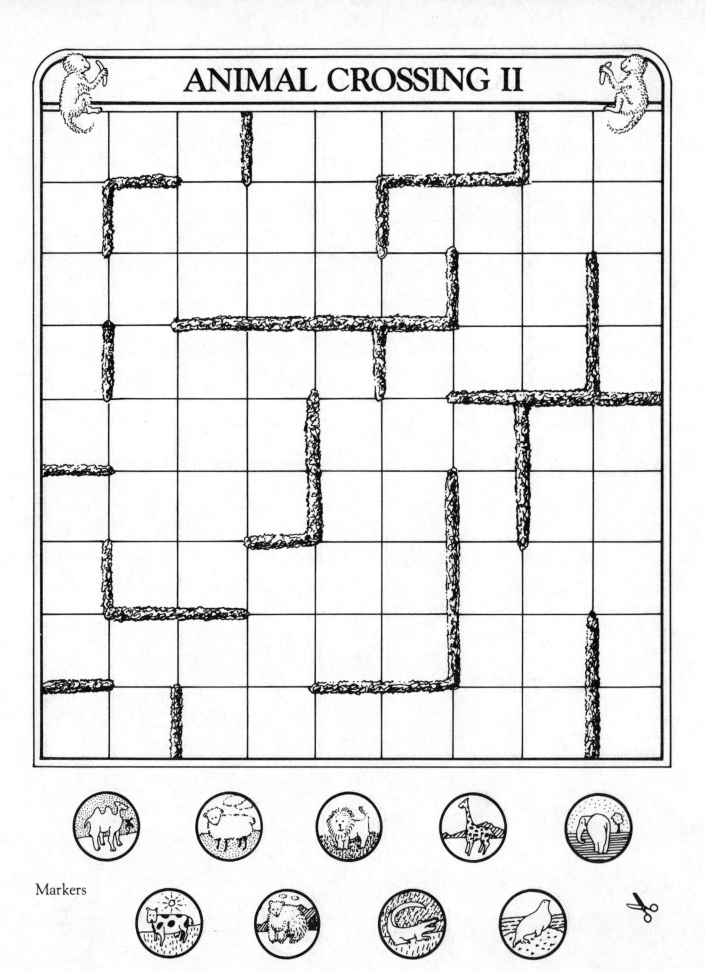

Markers

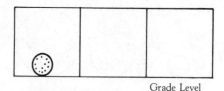

Why

To help children develop an understanding of the order of numbers and learn about the process of elimination

How

□ Choose a range of numbers for this activity that is appropriate for your child. For kindergarteners, 1 through 10 is good; first graders and older children can use 0 through 20 or higher. Each player needs a number line that covers his or her range.

□ The leader decides on a secret number from the number line.

□ The goal is for the players to guess the number in as few tries as possible.

□ The players take turns guessing.

□ The leader gives a clue response of "too large" or "too small" for each incorrect guess.

□ The players can place markers of one color on numbers that are too small and markers of another color on numbers that are too large. This will help them narrow the field of possible guesses.

□ The player who guesses the correct number is the next leader.

Grade Level

TOOLS

Number lines made from graph paper (see pages 79-82)

Markers to cover eliminated number

Number line

1	2	3	4	5	6	7	8	9	10

0	1	2	3	4	5	6	7	8	9	10	11	12	13	14

Balloon Ride

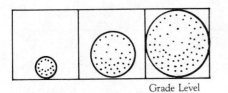

Grade Level

TOOLS

10 or more toothpicks
Balloon Ride board

A game for
2 players

Why

To practice problem-solving techniques by trying to figure out how to win a variation of the old Chinese game of NIM

How

☐ Tell the children a story about the hot air balloon coming to town. There is a contest to win a free ride. There are ten ropes holding the balloon to the ground. Two people take turns cutting the ropes. Each person may cut either **one** or **two** ropes. The person who cuts the last rope wins a free ride.

☐ Put out ten toothpicks on the Balloon Ride board to represent the ropes.

☐ Players take turns picking up one or two toothpicks at a time.

☐ No one is allowed to skip a turn.

☐ The person who takes the last one (or two) toothpicks wins the free ride.

▶ *When you and your child begin to see some patterns and possible strategies, see if together you can work out a way to win every time. (Hint: Start with a game that has just a few toothpicks. Who has the best chance of winning? Then add a few more toothpicks. This is called working backward.)* ◀

More Ideas

☐ After you think you have found a way to win, use a larger number of toothpicks—maybe twelve or nineteen.

☐ You can also change the number of toothpicks that can be picked up—try picking up one, two, or three toothpicks on each turn.

☐ You may even want to change the rules so that the person who has to pick up the last toothpick is the **loser** instead of the winner.

▶ *This activity develops intuitive understanding of subtraction or "take-away." If children can find a strategy to win, it will build a stronger number sense.* ◀

BALLOON RIDE

Arrange 10 toothpicks to connect balloon to ground

Target Addition

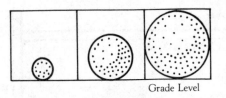

Grade Level

TOOLS

Game board (below)
Markers

A game for
2–4 players

Why

To provide practice in mental arithmetic and strategic planning

How

☐ Choose a target number between 25 and 55.

☐ Take turns placing a marker on one of the numbers on the board, each time announcing the total of the covered numbers.

☐ For example, if the first player covered a 4, the second a 3 and the third a 2, the sum would be 4+3+2 or 9. If the fourth player covered a 4, the total would be 9+4 or 13.

☐ Each square may be used only once.

☐ The first player to reach the target number **exactly** wins. If a player goes over the target number he or she is out.

5	5	5	5	5
4	4	4	4	4
3	3	3	3	3
2	2	2	2	2
1	1	1	1	1

Value of Words

Why

To practice mental arithmetic and estimation while problem-solving

How

- Assign values to the letters of the alphabet, as shown:
- Have each person in your family find the value of his or her first name.
- Add up the numbers without using paper and pencil if you can.
- What is the most expensive word each of you can find?
- Can you find a word worth exactly $50? $100?

More Ideas

- You and your child may want to make up different activities, such as:
  - Hold a week's contest to find the most expensive word.
  - Use penny values instead of dollars.
  - Find the difference between your first and last names.
  - Multiply the values instead of adding them.
  - Use fractional values, so that A=1/26, B=2/26, etc.

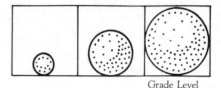

Grade Level

TOOLS

Pencil
Paper

A = $ 1	N = $14
B = $ 2	O = $15
C = $ 3	P = $16
D = $ 4	Q = $17
E = $ 5	R = $18
F = $ 6	S = $19
G = $ 7	T = $20
H = $ 8	U = $21
I = $ 9	V = $22
J = $10	W = $23
K = $11	X = $24
L = $12	Y = $25
M = $13	Z = $26

Bridges

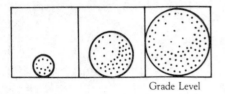

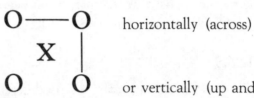

Grade Level

TOOLS

Bridges board (cover with clear plastic if possible)

Pencils or wash-off pens

*A game for
2 players*

Why

To practice planning ahead and keeping a watchful eye on opponents' moves

How

☐ There is an O player and an X player.

☐ The O player may connect any two adjacent O's (next to each other):

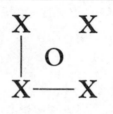

horizontally (across)

or vertically (up and down)

☐ The X player may connect any two adjacent X's horizontally or vertically.

$$X \qquad X$$
$$| \qquad O$$
$$X{-}{-}X$$

☐ Neither player may draw through a line that is already on the board.

☐ Players take turns connecting pairs of X's or O's.

☐ The O player wins if he or she makes a path from the top row of O's to the bottom; the X player wins if he or she makes a path from the left hand row to the right hand row of X's.

▸ *This activity is also good for spatial experience with vertical and horizontal relationships.* ◂

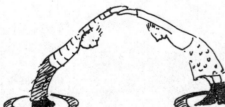

BRIDGES BOARD

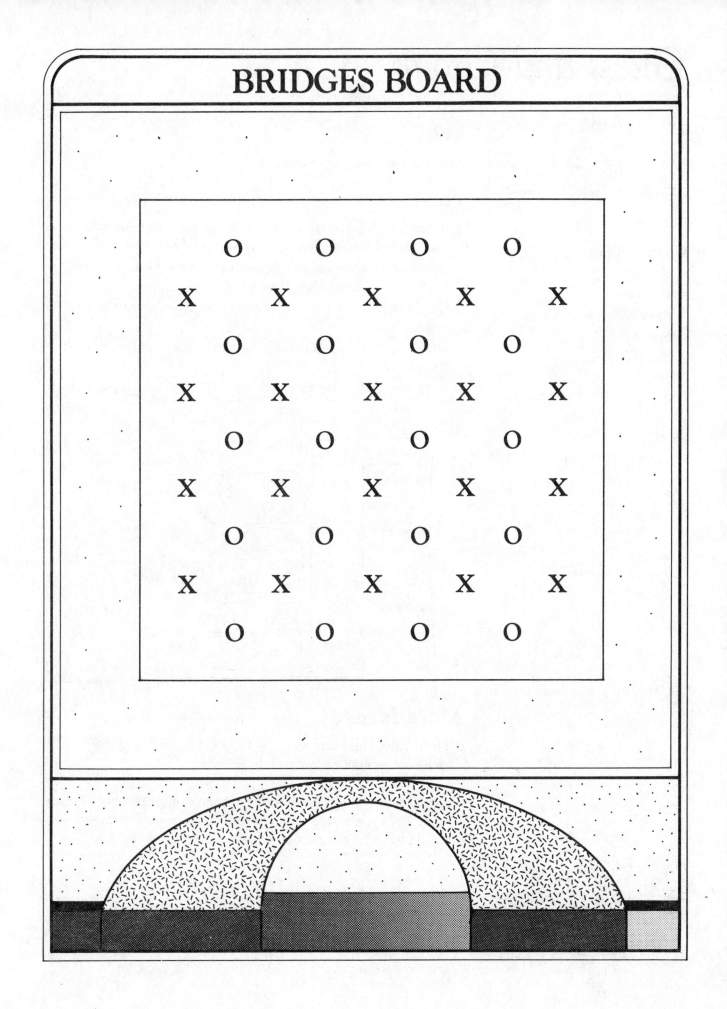

Guess and Group

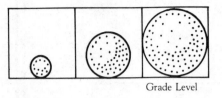

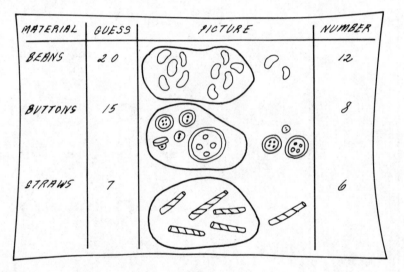

Grade Level

TOOLS

Containers of:
  Beans
  Buttons
  Small blocks
  Straws
  Toothpicks
  etc.

Why

To teach estimation and grouping

How

- ☐ Reach into the container of one of the materials and take all you can hold in one hand. Have your child also take a handful.
- ☐ Before you open your hands and look, guess how many you are holding. Write down your guesses.
- ☐ Count how many pieces you have. Count by groups of five or ten.
- ☐ Draw a picture of your count and write the number on the record sheet.
- ☐ Continue with other materials, as shown in the sample record sheet below.

MATERIAL	GUESS	PICTURE	NUMBER
BEANS	20		12
BUTTONS	15		8
STRAWS	7		6

More Ideas

- ☐ Have older children count by 3's, 4's, 6's, 7's, 8's, and 9's, to reinforce multiplication and division facts.

▶ *Grouping is a basic idea in developing the number concepts of place value and multiplication and division. Children need many grouping experiences with a variety of materials and different numbers.* ◀

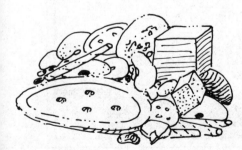

The Sum What Dice Game

Why

To practice basic addition facts and mental arithmetic

How

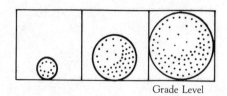

Grade Level

TOOLS

2 dice
Playing strips
Beans or other markers, or
Pencil and paper

A game for
2–4 players

- ☐ Give a playing strip to each player or have each person write out the numerals 1 through 9 on paper.

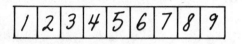

- ☐ Players take turns rolling two dice.
- ☐ On each turn the player may cover either the sum rolled on the dice or **any two numbers** that are still uncovered and that add to the sum rolled.

  For example, if a sum 9 is rolled first, the player may cover: 9, **or** 1 and 8, **or** 2 and 7, **or** 3 and 6, **or** 4 and 5.

- ☐ Later in the game if the sum of 9 is rolled again and the 5 is already covered, then the player cannot use the 4 and 5 combination and must play one of the other open possibilities.
- ☐ When a player cannot play, he or she is out and has a score of the sum of the **un**covered numbers.
- ☐ Play continues for everybody else until everyone is out.
- ☐ The last person to go out will not necessarily win; the person with the **lowest** score wins.

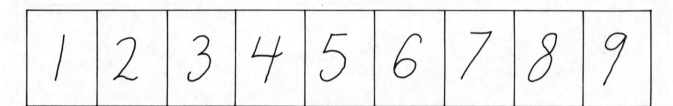

Playing strip

Paying the Price

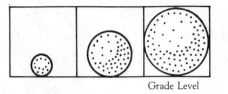

Grade Level

TOOLS

50 pennies
10 nickels
5 dimes
2 quarters
Coin board

Why

To become familiar with the value of coins and to practice making an organized list

How

☐ Help your child find how many different ways you could pay for **each of the items** using pennies, nickels, dimes, and/or quarters.

☐ For each way, put the coins on a separate row of the board below, then make a record of how many ways there were.

☐ For example, for a LOLLIPOP that costs five cents, there would be two ways:

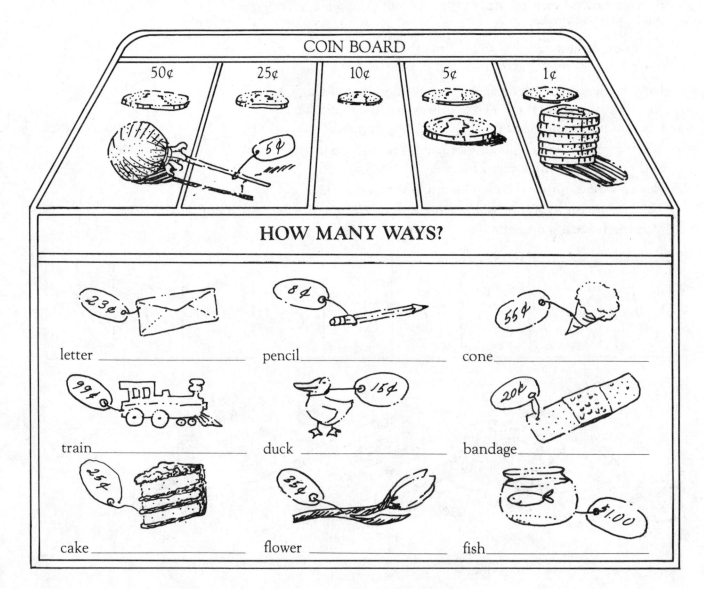

COIN BOARD

50¢ 25¢ 10¢ 5¢ 1¢

HOW MANY WAYS?

letter _____ 23¢

pencil _____ 8¢

cone _____ 55¢

train _____ 99¢

duck _____ 15¢

bandage _____ 20¢

cake _____ 25¢

flower _____ 35¢

fish _____ $1.00

COIN BOARD

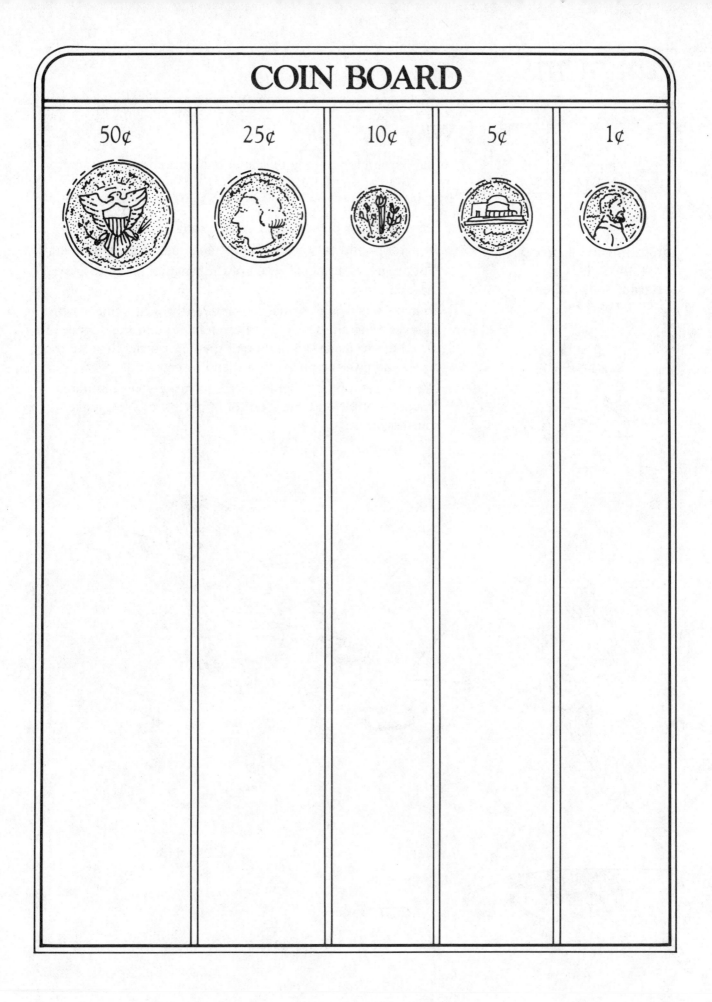

| 50¢ | 25¢ | 10¢ | 5¢ | 1¢ |

Tangrams

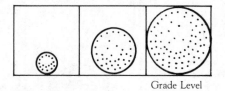

Grade Level

TOOLS

Tangram puzzle pieces
  (see page 42)

Tangram shape sheet
  (see page 43)

Why

To develop understanding of spatial relationships

How

☐ See page 42 for directions to make a tangram set.

☐ Help your child cut out the seven tangram pieces very carefully.

☐ Work together, using all seven of the tangram pieces to cover the bird shape.

☐ To make more puzzles, make an interesting arrangement with all seven tangram pieces. Trace carefully around the outside of the design to make the outline of the new puzzle. Remove the pieces and give the puzzle to a friend to try.

☐ Use all seven of the tangram pieces to make your first initial. Trace its outline and cut it out of bright paper. Now make your last initial.

BIRD SHAPE

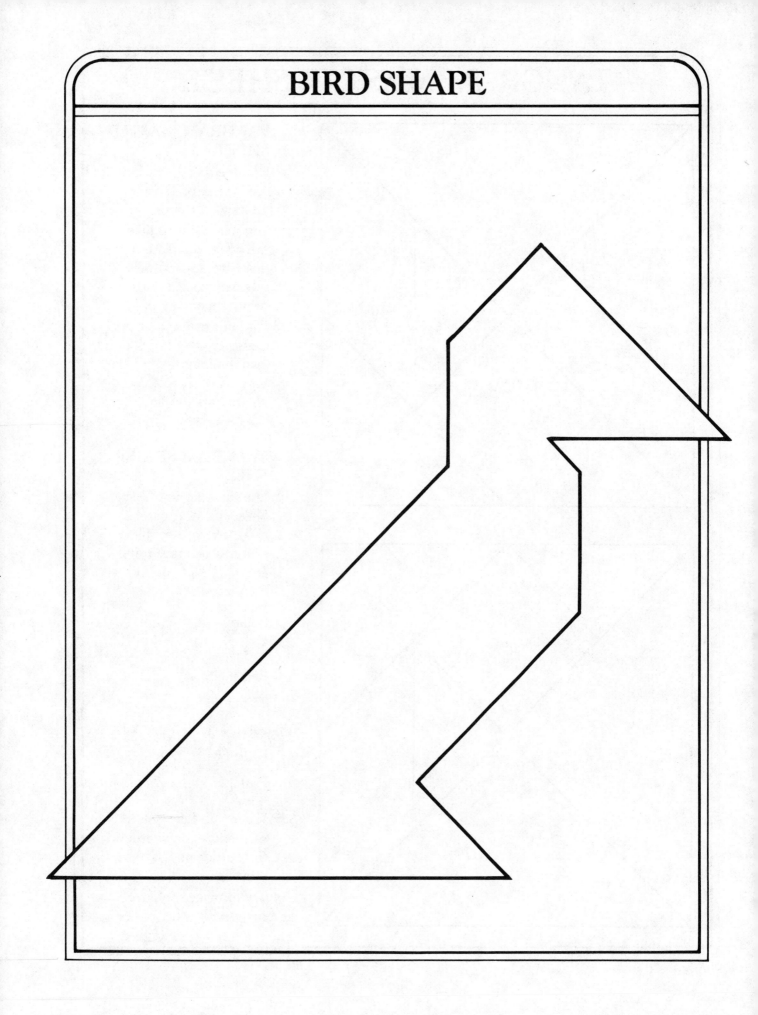

TANGRAM MASTER SHEET

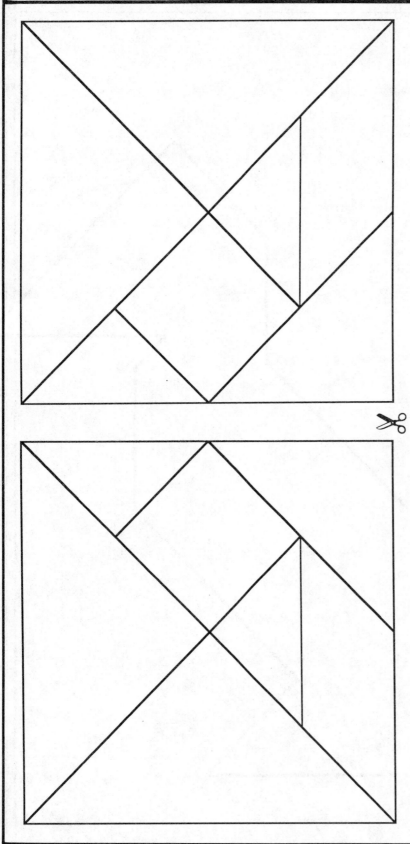

TANGRAM MASTER SHEET

☆ Each person should have one square of Tangram pieces to cut out.

☆ You may want to take the book to a place that makes copies and have this page copied onto heavy paper.

☆ Put your own initials on the seven pieces of your own Tangram shape. Save all of the pieces in an envelope, so you can use them later.

TANGRAM SHAPE SHEET

☆ Use the Tangram Shape Sheet (see next page) to make a record of shapes that can be made using 1, 2, 3, 4, 5, 6, or 7 of the pieces.

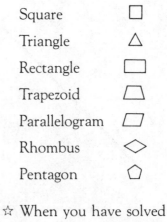

Square

Triangle

Rectangle

Trapezoid

Parallelogram

Rhombus

Pentagon

☆ When you have solved each shape, draw around the outside and save the drawing.

Try to make the same shape with other pieces.

TANGRAM SHAPE SHEET

Which shapes can you make with your tangram pieces? Draw a sketch of your solution.

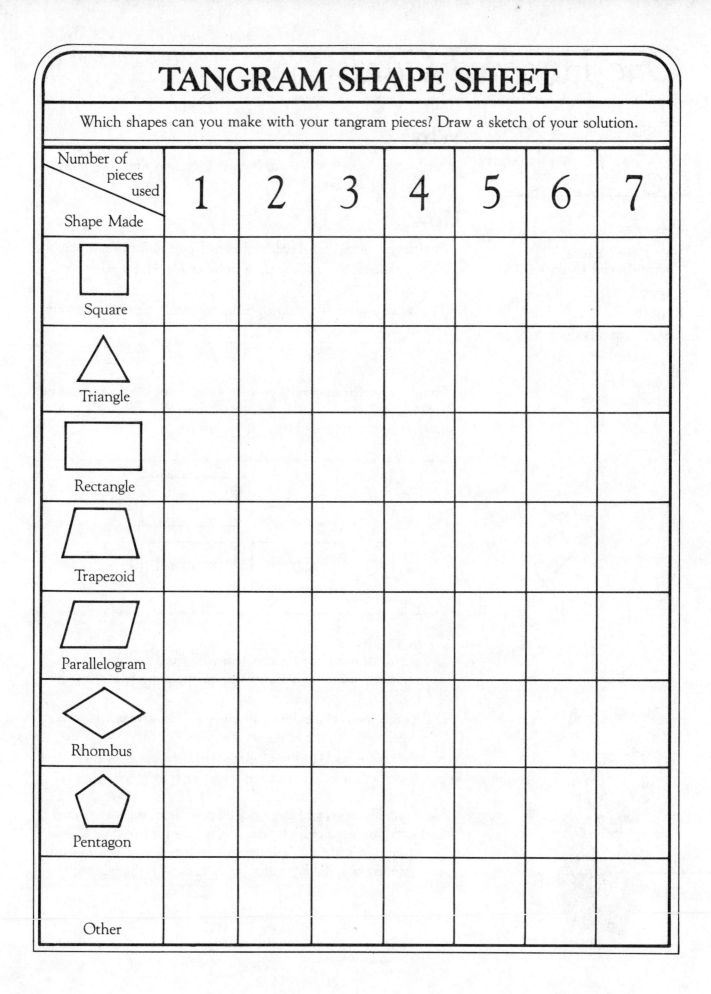

Number of pieces used / Shape Made	1	2	3	4	5	6	7
Square							
Triangle							
Rectangle							
Trapezoid							
Parallelogram							
Rhombus							
Pentagon							
Other							

One Hundred Cards

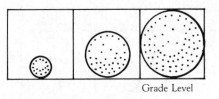

Grade Level

TOOLS

**100 3″×5″ cards
(numbered from 1-100)**
Paper
Pencils
List of answers (next page)

*A game for
2 or more people*

Why

To add a new twist for flashcards, which can help extend number sense beyond basic facts

How

☐ Shuffle the cards and distribute them face down to each player.

☐ Players take turns, holding up a number for all the other players.

☐ The other players write down as many matching multiplication problems as they can.

☐ To match, a problem must have the card's number as an answer or product.

For example, the matching problems for the number 24 would be 1×24, 2×12, 3×8, and 4×6. 24×1 counts the same as 1×24, 12×2 counts the same as 2×12, and so on.

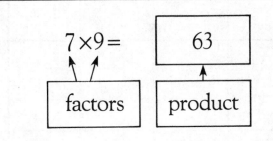

☐ To score, give:

  ☐ 1 point for each correct problem with both factors less than 11 or with 1 as a factor (for example, 6×5, 3×10, or 1×30);

  ☐ 2 points for each correct pair with at least one factor greater than 10 (for example, 2×15 or 3×21);

  ☐ 3 points for recognizing a prime number (no problems except one times the number itself, such as 1×17 or 1×37)

▶ *Have available a copy of the list of all possible multiplication problems with products from 1 to 100. Making the list would be a good project for an older child. The answers (or lists of problems to match each number) may also be written on the backs of the number cards.* ◀

FACTORS & PRODUCTS

$1 = 1 \times 1$	$24 = 1 \times 24$	$41 = 1 \times 41$	$58 = 1 \times 58$	$73 = 1 \times 73$	$88 = 1 \times 88$
$2 = 1 \times 2$	$\quad = 2 \times 12$	$42 = 1 \times 42$	$\quad = 2 \times 29$	$74 = 1 \times 74$	$\quad = 2 \times 44$
$3 = 1 \times 3$	$\quad = 3 \times 8$	$\quad = 2 \times 21$	$59 = 1 \times 59$	$\quad = 2 \times 37$	$\quad = 4 \times 22$
$4 = 1 \times 4$	$\quad = 4 \times 6$	$\quad = 3 \times 14$	$60 = 1 \times 60$	$75 = 1 \times 75$	$\quad = 8 \times 11$
$\quad = 2 \times 2$	$25 = 1 \times 25$	$\quad = 6 \times 7$	$\quad = 2 \times 30$	$\quad = 3 \times 25$	$89 = 1 \times 89$
$5 = 1 \times 5$	$\quad = 5 \times 5$	$43 = 1 \times 43$	$\quad = 3 \times 20$	$\quad = 5 \times 15$	$90 = 1 \times 90$
$6 = 1 \times 6$	$26 = 1 \times 26$	$44 = 1 \times 44$	$\quad = 4 \times 15$	$76 = 1 \times 76$	$\quad = 2 \times 45$
$\quad = 2 \times 3$	$\quad = 2 \times 13$	$\quad = 2 \times 22$	$\quad = 5 \times 12$	$\quad = 2 \times 38$	$\quad = 3 \times 30$
$7 = 1 \times 7$	$27 = 1 \times 27$	$\quad = 4 \times 11$	$\quad = 6 \times 10$	$\quad = 4 \times 19$	$\quad = 5 \times 18$
$8 = 1 \times 8$	$\quad = 3 \times 9$	$45 = 1 \times 45$	$61 = 1 \times 61$	$77 = 1 \times 77$	$\quad = 9 \times 10$
$\quad = 2 \times 4$	$28 = 1 \times 28$	$\quad = 3 \times 15$	$62 = 1 \times 62$	$\quad = 7 \times 11$	$\quad = 6 \times 15$
$9 = 1 \times 9$	$\quad = 2 \times 14$	$\quad = 5 \times 9$	$\quad = 2 \times 31$	$78 = 1 \times 78$	$91 = 1 \times 91$
$\quad = 3 \times 3$	$\quad = 4 \times 7$	$46 = 1 \times 46$	$63 = 1 \times 63$	$\quad = 2 \times 39$	$\quad = 7 \times 13$
$10 = 1 \times 10$	$29 = 1 \times 29$	$\quad = 2 \times 23$	$\quad = 3 \times 21$	$\quad = 3 \times 26$	$92 = 1 \times 92$
$\quad = 5 \times 2$	$30 = 1 \times 30$	$47 = 1 \times 47$	$\quad = 7 \times 9$	$\quad = 6 \times 13$	$\quad = 2 \times 46$
$11 = 1 \times 11$	$\quad = 2 \times 15$	$48 = 1 \times 48$	$64 = 1 \times 64$	$79 = 1 \times 79$	$\quad = 4 \times 23$
$12 = 1 \times 12$	$\quad = 3 \times 10$	$\quad = 2 \times 24$	$\quad = 2 \times 32$	$80 = 1 \times 80$	$93 = 1 \times 93$
$\quad = 2 \times 6$	$\quad = 5 \times 6$	$\quad = 3 \times 16$	$\quad = 4 \times 16$	$\quad = 2 \times 40$	$\quad = 3 \times 31$
$\quad = 3 \times 4$	$31 = 1 \times 31$	$\quad = 4 \times 12$	$\quad = 8 \times 8$	$\quad = 4 \times 20$	$94 = 1 \times 94$
$13 = 1 \times 13$	$32 = 1 \times 32$	$\quad = 6 \times 8$	$65 = 1 \times 65$	$\quad = 5 \times 16$	$\quad = 2 \times 47$
$14 = 1 \times 14$	$\quad = 2 \times 16$	$49 = 1 \times 49$	$\quad = 5 \times 13$	$\quad = 8 \times 10$	$95 = 1 \times 95$
$\quad = 2 \times 7$	$\quad = 4 \times 8$	$\quad = 7 \times 7$	$66 = 1 \times 66$	$81 = 1 \times 81$	$\quad = 5 \times 19$
$15 = 1 \times 15$	$33 = 1 \times 33$	$50 = 1 \times 50$	$\quad = 2 \times 33$	$\quad = 9 \times 9$	$96 = 1 \times 96$
$\quad = 3 \times 5$	$\quad = 3 \times 11$	$\quad = 2 \times 25$	$\quad = 3 \times 22$	$\quad = 3 \times 27$	$\quad = 2 \times 48$
$16 = 1 \times 16$	$34 = 1 \times 34$	$\quad = 5 \times 10$	$\quad = 6 \times 11$	$82 = 1 \times 82$	$\quad = 3 \times 32$
$\quad = 2 \times 8$	$\quad = 2 \times 17$	$51 = 1 \times 51$	$67 = 1 \times 67$	$\quad = 2 \times 41$	$\quad = 4 \times 24$
$\quad = 4 \times 4$	$35 = 1 \times 35$	$\quad = 3 \times 17$	$68 = 1 \times 68$	$83 = 1 \times 83$	$\quad = 6 \times 16$
$17 = 1 \times 17$	$\quad = 5 \times 7$	$52 = 1 \times 52$	$\quad = 2 \times 34$	$84 = 1 \times 84$	$\quad = 8 \times 12$
$18 = 1 \times 18$	$36 = 1 \times 36$	$\quad = 2 \times 26$	$\quad = 4 \times 17$	$\quad = 2 \times 42$	$97 = 1 \times 97$
$\quad = 2 \times 9$	$\quad = 2 \times 18$	$\quad = 4 \times 13$	$69 = 1 \times 69$	$\quad = 3 \times 28$	$98 = 1 \times 98$
$\quad = 3 \times 6$	$\quad = 3 \times 12$	$53 = 1 \times 53$	$\quad = 3 \times 23$	$\quad = 4 \times 21$	$\quad = 2 \times 49$
$19 = 1 \times 19$	$\quad = 4 \times 9$	$54 = 1 \times 54$	$70 = 1 \times 70$	$\quad = 7 \times 12$	$\quad = 7 \times 14$
$20 = 1 \times 20$	$\quad = 6 \times 6$	$\quad = 2 \times 27$	$\quad = 2 \times 35$	$\quad = 6 \times 14$	$99 = 1 \times 99$
$\quad = 2 \times 10$	$37 = 1 \times 37$	$\quad = 3 \times 18$	$\quad = 5 \times 14$	$85 = 1 \times 85$	$\quad = 3 \times 33$
$\quad = 4 \times 5$	$38 = 1 \times 38$	$\quad = 6 \times 9$	$\quad = 7 \times 10$	$\quad = 5 \times 17$	$\quad = 9 \times 11$
$21 = 1 \times 21$	$\quad = 2 \times 19$	$55 = 1 \times 55$	$71 = 1 \times 71$	$86 = 1 \times 86$	$100 = 1 \times 100$
$\quad = 3 \times 7$	$39 = 1 \times 39$	$\quad = 5 \times 11$	$72 = 1 \times 72$	$\quad = 2 \times 43$	$\quad = 2 \times 50$
$22 = 1 \times 22$	$\quad = 3 \times 13$	$56 = 1 \times 56$	$\quad = 2 \times 36$	$87 = 1 \times 87$	$\quad = 4 \times 25$
$\quad = 2 \times 11$	$40 = 1 \times 40$	$\quad = 2 \times 28$	$\quad = 3 \times 24$	$\quad = 3 \times 29$	$\quad = 5 \times 20$
$23 = 1 \times 23$	$\quad = 2 \times 20$	$\quad = 4 \times 14$	$\quad = 4 \times 18$		$\quad = 10 \times 10$
	$\quad = 4 \times 10$	$\quad = 7 \times 8$	$\quad = 6 \times 12$		
	$\quad = 5 \times 8$	$57 = 1 \times 7$	$\quad = 8 \times 9$		
		$\quad = 3 \times 19$			

Measure 15

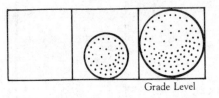

Grade Level

TOOLS

Metric ruler
Objects to measure
Record sheet

Why

To practice estimating and measuring with centimeters

How

- ☐ Make a record sheet like the one below.
- ☐ You and your child should estimate if your pencils are **shorter than** 15 centimeters (1.5 decimeters), **about** 15 cm long, or **longer than** 15 cm.
- ☐ Check the appropriate record sheet columns to indicate your estimates.
- ☐ Then measure the length of the pencils.
- ☐ Follow the same procedure for the other objects.
- ☐ Find more objects to estimate and measure. Are you both better at estimating now?

OBJECT	ESTIMATE			MEASUREMENT
	LESS THAN 15 cm.	ABOUT 15cm	GREATER THAN 15cm	
PENCIL	✓ DADS	✓ JOHNS		JOHNS - 12 cm DADS - 9 cm
BOOK				
TABLE				
FROM YOUR ELBOW TO YOUR WAIST				
THE LENGTH OF YOUR FOOT				

More Ideas

- ☐ To make the activity more challenging, only use objects that are close to 15 cm in length.
- ☐ Do the same activity for 12 inches (1 foot).

15 centimeter ruler

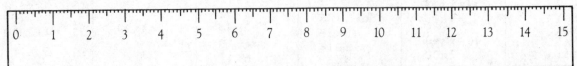

Nimble Calculator

Versions of the ancient game of NIM

Why

To develop strategies, use patterns as clues, practice addition and subtraction, and learn calculator operation skills

How

- ☐ See next page for game directions for all of the games.
- ☐ Play 7 UP with your child at least five times.
- ☐ After each game, let the loser decide whether to play first or second in the next game.
- ☐ Watch for the point in each game when it is clear who will win.
- ☐ Discuss any patterns found and ideas about winning strategies. Encourage your family to test theories with further play. It is important not to rush the playing time.
- ☐ Play each of the other games on the instruction sheet several times.
- ☐ Practice looking for patterns and strategies that will help you win.

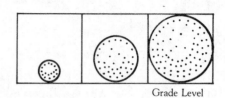

Grade Level

TOOLS

1 calculator for 2 people
Nimble Calculator
Instruction sheet
(next page)

NIMBLE CALCULATOR
Instruction Sheet

Games for 2 players and 1 calculator

7 UP

Clear the calculator so it reads 0. Taking turns, each player adds 1 or 2 into the calculator. The winner is the first person to reach 7. Going over 7 loses.

Start: 0
Add 1 or 2
Target: 7

11 Down

Clear the calculator and enter 11. Subtract 1 or 2 on each turn. Winner is person to reach 0.

Start: 11
Subtract 1 or 2
Target: 0

Now You're 21

Clear the calculator so it reads 0. Add 1, 2, 3, or 4 on each turn. The person to reach 21 wins.

Start: 0
Add 1-4
Target: 21

Travel Down 101

Enter 101 on the calculator. Each person in turn subtracts 1, 2, 3, 4, 5, 6, 7, 8, or 9 from the number in the display. Make the display read 0 and you win.

Start: 101
Subtract 1-9
Target: 0

Century

Start at 0 and add 1-9 on each turn until 100 is reached. This one has a twist.

Start: 0
Add 1-9
Target 100

2001

Enter 2001 in the calculator. Subtract 1–99 on each turn. The person to reach 0 wins.

Start: 2001
Subtract 1-99
Target: 0

Shopping Spree

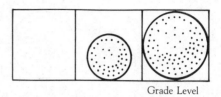

Grade Level

TOOLS

Calculator

Why

To learn calculator skills

► Many calculators have memory keys that allow you to store and retrieve intermediate results of a larger calculation. The five memory features (though they may have different names on different calculators) are:

Memory Clear MC

—clears the memory of anything previously stored there

Memory Recall MR

—displays the number stored in memory

Memory Add M+

—adds the number on the display to the number stored in the memory

Memory Minus M–

—subtracts the number on the display from the number stored in the memory

A reminder light tells you that something is in memory. ◄

How

With your child, experiment with a calculator to see how it handles these memory features. For more information see the Chapter on Estimation and Calculators. Following is a shopping spree story problem you might try to solve, using your calculator without paper or pencil. (And remember, math **should** be fun!)

One afternoon, when the old Coca Cola thermometer read 120°, Aunt Bebe noticed that her battery was on the blink. So she beat a path to the hardware store, where she bought 80 battery packs at $1.80 a pack. Bowing to impulsive pressure, she also bought seven beautiful bright begonia bouquets at $.49 apiece and eight computer disks. (They were on special at two for $13.56.) Naturally she also bought a fan to cool herself and her computer for $27.55. What could the bulging total fee be for Bebe's shopping spree?

WARNING: Here is the answer in code so you don't accidentally look at it before you want to. To break the code, look at it in a mirror and add $99.99 to what you see.

Orderly Operations

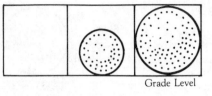

Grade Level

TOOLS

Calculators

Why

To develop understanding of calculators and how they work

How

- □ Work together on the problems below.
- □ Study the first problem to decide if entering the numbers and operational symbols in the order shown will give the answer 22.
- □ Then enter the problem into the calculator to check.
- □ Continue with the other problems, estimating first each time.

PROBLEMS:

$3 \times 6 + 4 = 22$

$5 \times (7-3) + 4 = 24$

$4 \times (12 \div (6-3)) = 16$

$17 - 3 \times 5 = 2$

$80 \times 3/5 = 48$

$80 \div 3/5 = 133.33$

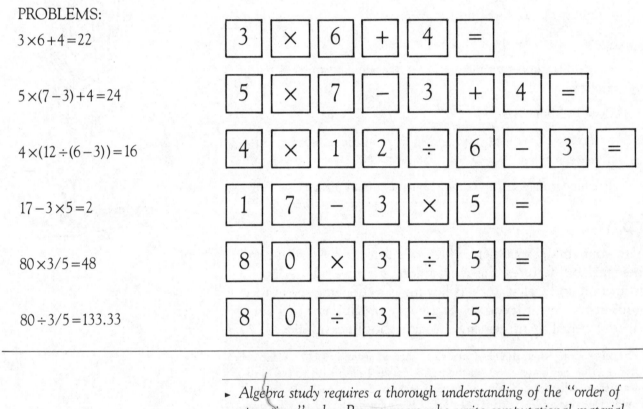

> Algebra study requires a thorough understanding of the "order of operations" rules. Programmers who write computational material for computers must have good knowledge of these rules. ◄

Chapter III

WORD PROBLEMS and LOGICAL REASONING

WORD PROBLEMS AND LOGICAL THINKING

Parents often ask us how to help their children with the word problems and logic taught in schools. This chapter contains a number of activities that involve logical thinking, along with some ideas for you to use in helping your children extend their repertoire of ways to tackle word problems.

TOOL KIT

Sorting objects, such as buttons, bottlecaps, plastic bread tags, seeds, beans, leaves, and so on

Large pieces of paper

Pens and pencils

Graph paper (see page 79-82)

5″×8″ cards

Dice

WORD PROBLEMS

If word problems have created a few tears and tearing of hair in your household, try some new techniques, such as using manipulatives, drawing pictures, breaking the problem into parts, guessing at the answer, or reducing the numbers to smaller ones.

You will be surprised at the power these techniques will provide over some really mean problems. And if one method doesn't seem to be working, try another. The practice is great. The examples here will give you some ideas about how to apply the strategies to many other problems.

Use manipulative materials to represent the parts of the problem.

> One day George was running home as fast as he could with 10 walnuts in his pocket. Out fell 2 walnuts, but then he found 3 more and put them in his pocket with the others. Then he lost 4 more and found another. How many did he have when he got home?

This is a tough problem for young children, but if you give them some beans or blocks (or walnuts), it becomes easy. First set out the original ten walnuts. Then read the phrase "out fell 2 walnuts," asking "Do we add or take away?" Your child will most likely say to take away two of the walnuts. The next phrase says "he found 3 more," which means adding three walnuts. And so on until the problem has been finished, step by step.

Draw a picture of the situation. This is usually a sure winner!

> Alice had 4 more cups of tea than the rabbit. Together they had 10 cups of tea. How many cups did each have?

Here is the rabbit's tea, but we don't know how many cups:

Here is Alice's, the same as the rabbit's, with four extra cups:

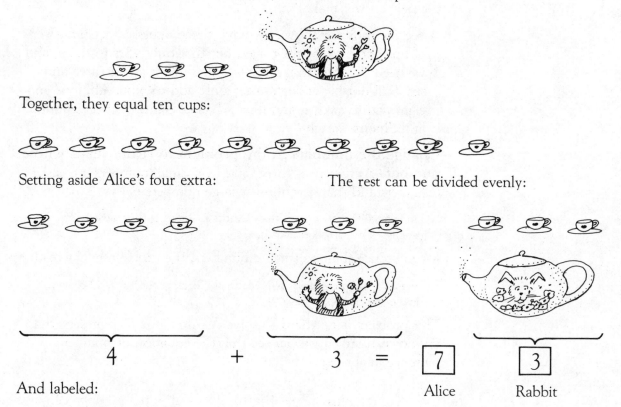

Together, they equal ten cups:

Setting aside Alice's four extra: The rest can be divided evenly:

$$4 \quad + \quad 3 \quad = \quad \boxed{7} \quad \boxed{3}$$

And labeled: Alice   Rabbit

Once you and your children have been drawing pictures for a while, they will probably become more like stick figures or diagrams, to shorten the problem-solving time, but do enjoy drawing the pictures as long as you like!

Break the problem down into smaller parts. Read the first sentence of the problem; stop and consider what it means. Maybe even do the calculations indicated. Then read the next sentence and stop again. Continue until you have read the whole problem, considering each part carefully. Often this is enough to clarify what needs to be done.

> Michael is ½ as old as Sharon.
> Together their ages are 12. How old is each?

☐ Upon reading and discussing the first sentence, you can see that Sharon is older, in fact is twice as old as Michael. If Michael is one year old, Sharon will be two.

☐ Reading the second sentence should lead to thinking of older ages, since if Michael is one and Sharon is two their ages together are only three.

☐ By this point, a **process** to find the solution is clear, and the rest is easy.

Guess at an answer then put the guess into the problem and see whether it makes sense. If not, guess again, and see whether the new answer makes sense.

☐ See the problem above. We have already guessed at an answer of one and two for their ages, but that only adds to three, and we need two ages that add to twelve. What about three and six? Still not big enough—they only add to nine. Ah! four and eight add to twelve, and four is half of eight. Michael is four and Sharon is eight!

Change the numbers in the problem to really small whole numbers such as two, three, and five. Small numbers are easier to understand than fractions or large numbers.

> Linda has 695 more pencils than Matthew. Together they have 4,725 pencils. How many does each have?

If you reduce these to smaller numbers, the problem might read:

> Linda has 4 more pencils than Matthew. Together they have 10 pencils. How many does each have?

This problem is not hard to solve. Using a box to represent the unknown number, we can see that the equation or number sentence might be:

$$\square \quad \text{plus} \quad \square + 4 = 10 \quad \text{or} \quad \square + (\square + 4) = 10$$

Matthew's	Linda's	$\square + \square + 4 = 10$
pencils	pencils	$\square + \square = 10 - 4$
		$\square + \square = 6$
		$\square = 3$

To check the answer, use ten real pencils, or draw a picture. Putting the bigger numbers back into the sentence give us:

$$\square + (\square + 695) = 4,725$$

And a little arithmetic on our calculator tells us that the number 2,015 goes into each box, meaning that Matthew has 2,015 pens and Linda has 2,710.

We hope you and your family will enjoy trying out these techniques, and perhaps will invent some of your own. One of the most effective strategies we know of is to let the problem sit for a day or two, or sleep on it! This is a paradoxical part of the admonition to "stick with it," because sometimes stepping back from the problem allows one to see that another path to solution does exist. If we work too long on a problem without getting anywhere, we are probably going in the same wrong direction over and over.

There are a number of books listed in the Resource List on page 311 with wonderful problems for you to try. Just remember to give yourselves plenty of time, to try several different techniques, and to enjoy the challenge.

LOGICAL REASONING

A natural extension of word problems is a process mathematicians call logical reasoning.

Jill and Jack had brought ten buckets of water down from the well, and they were about to go and get another. Jill said to Jack, "That doesn't make sense—the tank where we are putting the water is almost full and won't hold another bucketful!" Jack replied, "That's logical! I'm glad you saved us another trip!"

Joe and Jennie were storing groceries on the shelf. First they put all of the canned goods on one shelf. Then they put the bags of brown sugar and white sugar together on another shelf. When Jennie started to put the bag of flour with the canned goods, Joe said it was more logical to put it with the bags of sugar.

Logical thinking is simply making good sense out of something, usually in an organized way. It includes sorting things by some characteristics (for example, whether things are in bags or cans) or thinking ahead about what the results of an action (going after one more bucket of water) might be.

Most mathematics, including logical thinking, cannot be learned by hearing a good lecture. Logical thinking requires individual experimentation with the world. Children clarify and strengthen their reasoning abilities by talking about strategies. The home is an ideal place for youngsters to practice explaining how they **think** they think. In school, teachers don't have the time to listen extensively to how each child figures out each logical dilemma. In fact, if teachers did absolutely nothing but listen to children, they would be able to spend only ten minutes with each child during any day.

Why is this verbalization so important? Often a child can solve an easy problem but is at a total loss about a similar problem with "uglier" numbers (like $4\frac{3}{8}$ instead of 2). Encourage her to talk about how she solved the easy problem and she can then use the same process on the more difficult one. By talking it out, a child becomes conscious of a strategy.

This awareness has two benefits. First, the strategy becomes transferable—it can be used in other situations (with encouragement and practice, of course). Second, and perhaps most important, the child learns that, when faced with an intellectual dilemma, she should search for strategies—something can be done; she doesn't have to wait for a solution to miraculously appear; the abstract and concrete world can by played with, jiggled, turned upside down, explored, and comprehended.

But what about the child who doesn't have a logical mind? One of the most pervasive and destructive myths about mathematics is that many children are simply not equipped to learn it. Not everyone is capable of inventing calculus or of creating a mathematics model that explains the movement of asteroids in space, but nearly every one of us can learn and enjoy precollege mathematics. Special care should be taken not to cheat ourselves or our children out of developing logical reasoning skills because we think it will be impossible or hard.

In mathematics, there is both formal and informal logic. The formal kind of logic of mathematicians includes simple, but sometimes tricky, words. Using the cans and bags above, the flour **is** in the bag category and **is not** in the can category. A bunch of bananas is **not** in the bag **or** the can category. Students need a great deal of practice with using these words carefully to prepare for advanced mathematics and logic.

Venn Diagrams are a helpful way of learning to sort things into categories. While you are working with the Venn Diagrams in this Chapter, pay special attention to the "not" categories. These are like subtraction and division—often harder to use than addition and multiplication.

Guessing, estimating, and predicting are another vital part of logical thinking. In Rainbow Logic, children ask for some information from the leader. Based on that information, they should be able to ask for the next bit of information with "educated guessing," having given the matter some thought. Looking ahead to anticipate the next move or the results of an action is an important skill for all mathematics, as well as for living!

Some of these activities are games or puzzles that don't use numbers at all. Most of the games can be played repeatedly and deeper understanding of the strategies will occur with each playing. Discussing the games after you've played them is important to developing strategy skills, but too much discussion too soon will make the games deadly tasks. Children should have enough time just playing around to build their intuitive strategy-making abilities. Only after a lot of exploration is it a good idea to begin talking about "What would happen if I moved here?" "Who will win from this position?" or "What would be a good strategy for me to try now?"

When you have done the activities from this chapter, look through the other games on your family's game shelf, such as dominoes or checkers. Play the games and see whether some of the logic or strategies of the game become more clear when you and your child talk together about them. Some games, of course, are based on luck alone, but most are perfectly logical!

We hope that you and your family will become collectors of good logic games and good problems that involve logical thinking.

Sorting and Classifying

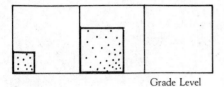

Grade Level

TOOLS

Buttons
Lids
Bottlecaps
Plastic bread tags
Seeds
Other objects to sort

Why

To develop the ability to observe similarities and differences and to practice language skills relating abstract ideas to the real world

How

☐ Give your child or children a collection of objects, say buttons, to sort.

☐ After the objects have been sorted, ask the children to explain their rules for sorting the buttons. Put out some more buttons to be sorted according to those rules.

☐ Children may choose a great variety of rules, such as red buttons, blue buttons, round buttons, square buttons, buttons with two holes, or four holes, and so on.

☐ Have the children discuss whether there is another way to sort, such as by large and small buttons, or gold buttons and not gold buttons.

☐ If children do not by themselves choose to sort using a concept of "not," sort them yourself, saying, for example, "These are green," "What can I say about these?" and "These are **not** green." Then ask the children to find another way to sort the buttons using the ideas of "not."

More Ideas

Make circles of string. Sort the objects into the circles to prepare for the Venn Diagram activity on the next page. Use two circles, or have a single circle with a rule for what goes inside and what does not go inside.

Venn Diagrams

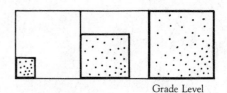

Grade Level

Why

To develop a logical way of sorting and classifying

How

► *Venn diagrams are usually drawn with circles and labels to indicate what belongs inside the circles.* ◄

TOOLS

Large paper
Pen or pencil

☐ This activity will involve making Venn diagrams for friends to sign or initial, and can be done with a class or over a longer period of time at home.

☐ These sketches illustrate various Venn diagrams and the kind of sorting statements that might be used with them:

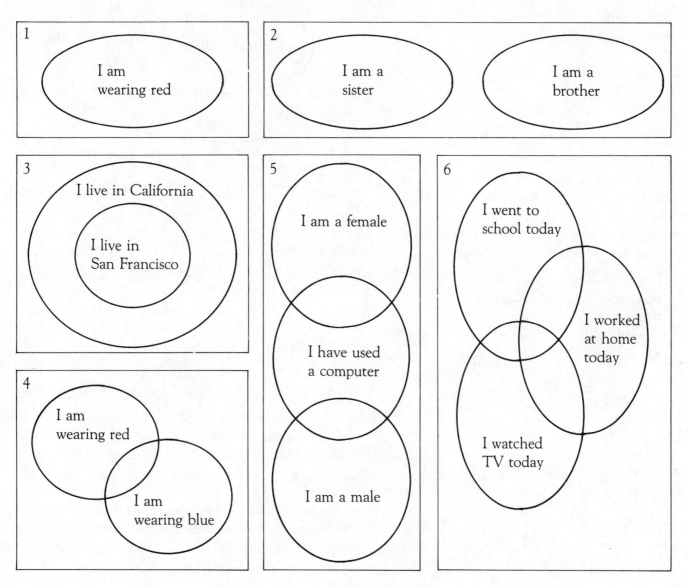

□ Have each person make a Venn Diagram on a large piece of paper. Choose any characteristics that some friends will have and some will not have.

□ Each friend signs his or her name in the circle or circles with the statements that are true for him or her. If no statement is true for a name, it goes outside all circles.

□ When there are overlapping circles, care must be taken to place the name in the appropriate sections of the diagrams.

□ **Some examples:**

In example A, if you live in Los Angeles, California, your name goes in the "I live in California" circle, but outside the "I live in San Francisco."

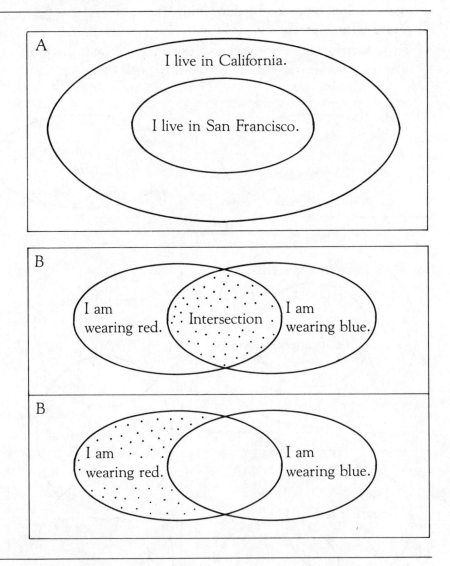

In B, if you are wearing both red and blue, your name has to go in both circles—in the central overlap area of the diagram, called the intersection.

If you are wearing red, but not blue, your name would go in the section of the diagram that is inside only the "I am wearing red" circle.

A

I live in California.

I live in San Francisco.

B

I am wearing red. · Intersection · I am wearing blue.

B

I am wearing red. I am wearing blue.

► *The diagrams are named after an Englishman nammed John Venn, who lived until 1923 and who made these symbolic drawings popular. This kind of logic is very important in science and advanced mathematics. It is also used in the design of computer circuits.* ◄

Two-Dimensional NIM

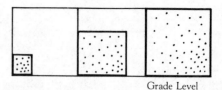

Grade Level

Why

To practice spatial reasoning and logical thinking

How

☐ Use the 3×6 rectangular grid below.

☐ Players take turns putting markers on one or two squares at a time.

☐ If two squares are covered, they must be fully connected on a side:

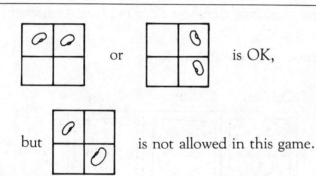

TOOLS

Graph paper grid
(see pages 79-82)

Two kinds of markers
(beans, bottle caps, etc.)

*A Nim game
for 2 players*

☐ No one can skip a turn.

☐ The person who fills in the last square (or the last two squares) wins.

More Ideas

☐ Play on a larger grid.

☐ Allow one, two, or three squares to be covered as long as they are fully connected on a side.

☐ Change the rules so that the loser is the one who covers the last square. (And **still** no skipping turns!)

Rainbow Logic

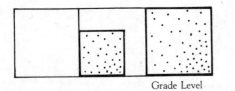
Grade Level

TOOLS

Colored paper squares for each player (4 each of 4 colors)

3×3 and 4×4 grids

*A game for
2 or more players*

Patterns like

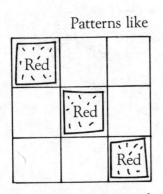

and

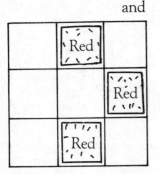

are not allowed.

Why

To practice deductive thinking and spatial reasoning

How

□ For the first game, a parent should be leader.

□ After the first game, any player may become leader.

□ The leader prepares a secret 3×3 color grid, using three squares of each color.

□ All of the squares of the same color must be connected by at least one full side.

For example, a secret grid might be

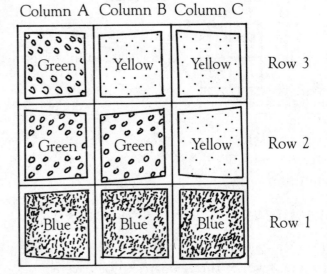

□ Clues are given in the following manner:
  □ Players ask for the colors in a row or a column (rows are horizontal, columns are vertical.)
  □ The leader gives the colors, but not necessarily in order.

□ Allow time for the players to discuss what they have learned after each guess before giving a new clue.

□ The goal is for the players to be able to give the location of all colors on the grid after as few questions as possible. Each player should use a grid and colored paper squares to keep track of the clues. Squares may be put **beside** the row or column until exact places are determined.

□ Let each person be the leader for two games, then let a new person lead, until all have had a chance to lead.

□ When everybody is familiar with the game, or for older students, play using a 4×4 grid, with the same rules.

More Ideas

☐ Either before beginning the game, or after you have played, talk about how many different possible arrangements there are for the three colors. See Pentasquares (page 188) for more discussion.

☐ For younger children, try a 2×2 grid, or give them the color information in order, so that they can put the colors onto the grid immediately.

	Column A	Column B	Column C	Column D	
					Row 4
					Row 3
					Row 2
					Row 1

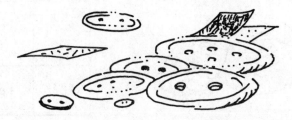

Pico, Fermi, Bagels

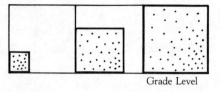

Grade Level

TOOLS

Pencil
Paper

Why

To practice making deductions by the process of elimination and to reinforce the concept of place value

How

☐ The goal is to guess the leader's secret three-digit number.

☐ The digits in the secret number must be all different; that is, 121, 442, and 777 are not allowed.

☐ The leader will give clue responses to each guess as follows:

BAGELS means **none** of the digits is correct

PICO means **one** of the digits is correct, but it is in the wrong place

FERMI means **one** of the digits is correct and in the correct place.

☐ This sample game will illustrate how the clues are given. *(The secret number is 427.)*

Guess	Response	Comments
109	Bagels	1, 0, and 9 are eliminated from all places
123	Fermi	Only one digit is correct and in the correct place
145	Pico	One of the digits is correct but in the wrong place
265	Pico	Same
353	Bagels	3 and 5 are now eliminated from all the places
426	Fermi Fermi	Two numbers are correct and in the correct spot
427	Fermi Fermi Fermi	All correct!!

☐ Players should keep a record of the guesses and leader's responses.

☐ If you are the leader, write your number on a slip of paper to refer to as you give the clues.

More Ideas

☐ Allow repeated digits.

☐ Play with more than three digits.

☐ Play with letters that form three-, four-, or five-letter words.

▸ **Pico** *is a metric prefix meaning one trillionth or* 10^{-12}.
Fermi *was a famous nuclear physicist.*
Bagel *is a hard roll with a center hole, like a zero.* ◂

Ten Card Arrangement

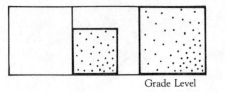

TOOLS

3×5 cards
Pens

Why

To use a complex but logical number series to solve a problem

How

☐ Give each pair of people (usually parent and child) a set of ten 3×5 cards.

☐ Have the numerals 1 through 10 written on the cards:

The Challenge

☐ Arrange the set of cards so that when you turn them face down in a stack the following happens as you turn each card over:

☐ The first (top) card is turned face up (to begin a new stack) and is a 1.

☐ The second card is moved to the bottom of the old stack, still face down.

☐ The third card is turned face up to the new stack and is a 2.

☐ The fourth card is moved to the bottom of the old stack, face down.

☐ The fifth card is turned face up and is a 3.

☐ The sixth card is moved to the bottom, face down.

☐ The seventh card is turned face up and is a 4.

☐ The eighth card...

...And so on until all of the cards have been turned face up, and the resulting new stack is in reverse order, with the 10 on top and the 1 on the bottom.

▸ *Some possible solution strategies are shown at the end of this chapter, on page 71. Please don't look until you have worked on the problem for at least 30 minutes. Try several strategies before looking.* ◂

Tax Collector

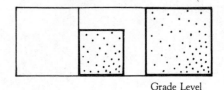

Why

To practice recognizing and solving problems that involve multiplication factors

The Problem

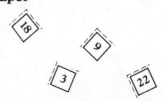

Grade Level

TOOLS

Paycheck squares, 1 through 24

Tax Collector Board (see page 69)

Pencil

Paper

- ☐ Your family is about to meet the tax collector! Your goal, of course, is to end the game keeping more money than the tax collector can get from you.
- ☐ The tax collector **must** however, receive something each time the taxpayer takes a paycheck. Payment is made in the form of **factors** of the taxpayer's check.

▶ *When two numbers are multiplied together, the answer is a* **product.** *The two numbers are* **factors.** *The multiplication problems for 16 are 1×16, 2×8, and 4×4, so the factors of 16 are 1, 2, 4, 8, and 16.*

▶ *A* **prime** *number has only two factors—itself and 1.*

▶ *You may want to review or list the factors of all the numbers before playing the first time (see page 45 for list).* ◀

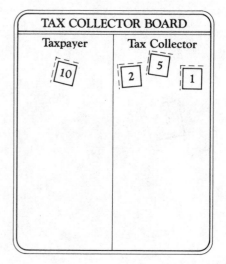

To Play

- ☐ Play the first game for practice, using only the numbers 1 through 12.
- ☐ Put out the twelve paycheck squares across the top of the Tax Collector Board.
- ☐ Pick a paycheck for the tax**payer,** and put it on that side of the board.
- ☐ Give the tax collector all the factors of that paycheck (or number).
- ☐ The tax collector will always get all of the paycheck squares that are factors of the taxpayer's check and that are still available.
- ☐ Once a number has been used, it may not be used again until the next game.
- ☐ Since **1** is a factor of every number, the tax collector will get **1** from the first paycheck that is chosen, along with any other factors.
- ☐ Continue choosing paychecks for the taxpayer and paying the tax collector until there are no paychecks left that have factors.
- ☐ If there are no factors left for a particular paycheck, the tax collector gets that check.

□ When there are no paychecks left that have factors available, the tax collector gets the rest.

□ Add the taxpayer's total and the tax collector's total to see who has the most.

□ Here is a sample game:

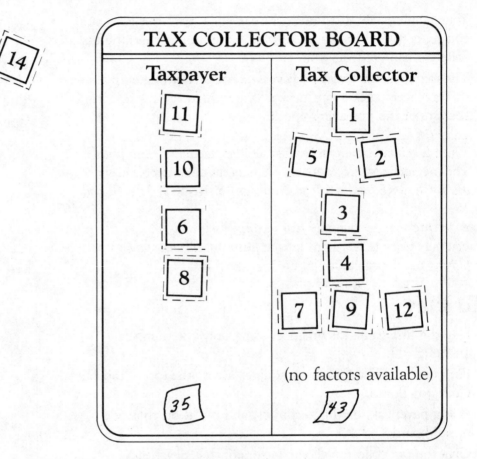

(Well, the tax collector won *that* time!)

□ Now that you know how to play the game, try it with all 24 of the paychecks. Be sure to work together and talk about why you should choose certain paychecks. Plan ahead. See if you can keep improving your score.

More Ideas

□ For students who are just learning multiplication facts, play several games with only twelve paychecks.

□ To make the game harder, play with more paychecks—up to 31 or maybe even 50!

TAX COLLECTOR BOARD

Taxpayer	Tax Collector

Paycheck Numbers:

1	2	3	4	5	6	7	8
9	10	11	12	13	14	15	16
17	18	19	20	21	22	23	24

Save Twenty

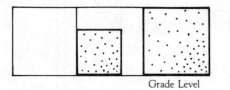

TOOLS

5 dice for each pair, or spinners (see page 154)

Scratch paper

*A game for
2 players*

Why

To build understanding of probability and to provide practice with strategy development, averaging, and addition

How

☐ In each game, five dice are rolled to try to create a sum close to, but not greater than, 20.

☐ A sum larger than 20 gives the player a score of zero.

☐ Players try to achieve the highest total score for ten games.

☐ Each game has **four** rounds.

☐ A player rolls all four rounds before giving the dice to another player.

☐ On the first round, five dice are rolled. For example:

☐ The player may "save" from none to all of the dice to make up her or his game score.

☐ Any dice not saved on the first round must be rolled again for round two. Again, none or all of the newly rolled dice may be saved.

☐ Continue this way through round four.

☐ On the fourth roll, all dice that are left must be used to make the final score.

☐ Note: Any dice saved cannot be rolled again in that game.

☐ Here is a sample game:

ROUND	ROLLED	SAVED	TOTAL	
1	⚅ ⚁ ⚀ ⚄ ⚀	⚅ ⚁ ⚃	12	TOTAL SCORE 18
2	⚀ ⚁	NONE	0	
3	⚄ ⚄	⚄	5	
4	⚀	⚀	1	

☐ After both players have played a game, they record their scores. After ten games, average the scores. The player with the highest average is the lucky winner.

More Ideas

Instead of pairs, play with teams. Talk to each other about your reasons for saving dice. Try to keep improving your team scores.

Ten Card Arrangement—solution strategies

(See page 66 for Ten Card Arrangement instructions)

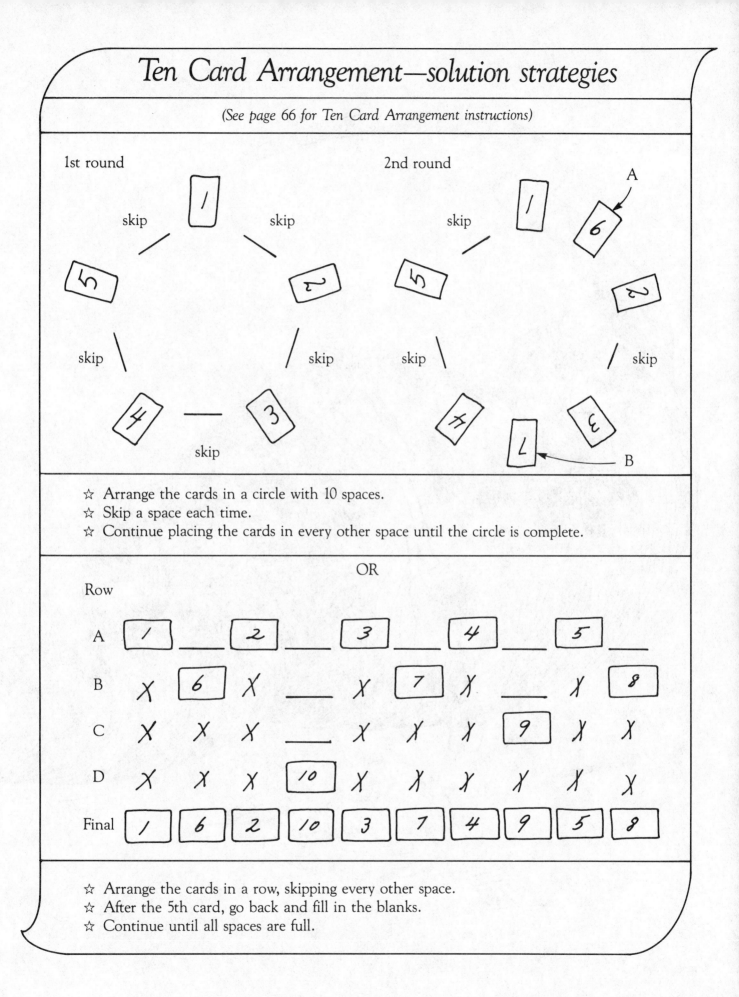

1st round

2nd round

☆ Arrange the cards in a circle with 10 spaces.
☆ Skip a space each time.
☆ Continue placing the cards in every other space until the circle is complete.

OR

Row

A | 1 | __ | 2 | __ | 3 | __ | 4 | __ | 5 | __
B | X | 6 | X | __ | X | 7 | X | __ | X | 8
C | X | X | X | __ | X | X | X | 9 | X | X
D | X | X | X | 10 | X | X | X | X | X | X

Final | 1 | 6 | 2 | 10 | 3 | 7 | 4 | 9 | 5 | 8

☆ Arrange the cards in a row, skipping every other space.
☆ After the 5th card, go back and fill in the blanks.
☆ Continue until all spaces are full.

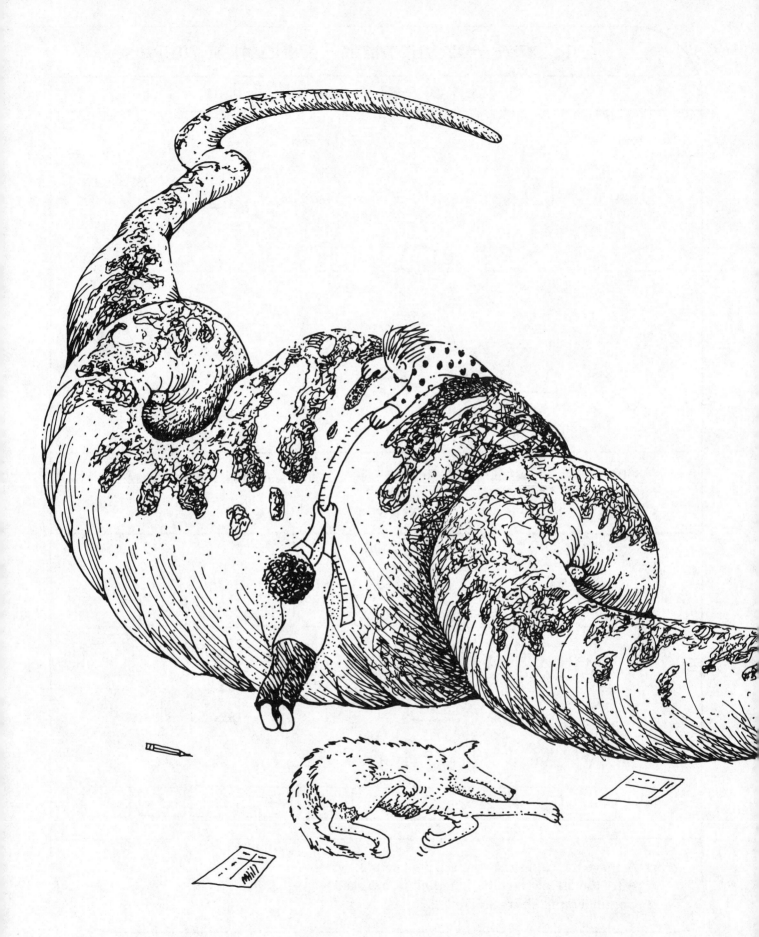

Chapter IV

MEASUREMENT

MEASUREMENT

TOOL KIT

Toothpicks—1 box

Paperclips—1 box

Beans—1 bag

String—1 ball

Adding machine tape—
   1 roll

Masking tape—1 small roll

Drawing paper

Tiles or one-inch paper
   squares—50 to 100

Graph paper—
   (see page 79-82)

Scissors

Pencils

Coloring pens or crayons

Ruler with markings to
   1/16 inch

Centimeter ruler

Meter stick and yardstick

Measuring cup

Liquid containers of
   various sizes

Assorted lids

Small boxes

Sugar cubes or small blocks

Balance scale

The ability to make use of the tools of measurement is one of the most important mathematical skills needed in daily life and on the job. The best way to develop these skills is by hands-on experience.

Our children learn how to make measurements of length, area, volume, capacity, weight, and temperature in elementary school. The table on page 78 gives a summary of these and the types of measurements and the units we might use to record measurements. Later, in high school, in subjects like algebra and physics, students will learn about other measurements such as velocity, acceleration, power, and energy.

Young children need to repeat their first measurement activities many times to acquire understanding of these concepts. For example, four- and five-year-olds will usually agree that the two pens in the first sketch are the same length, but if one pen is moved to the left or turned at right angles, they will often say that one pen is longer than the other.

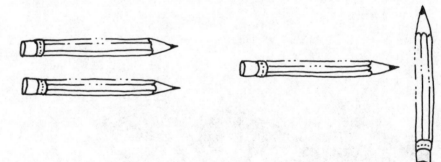

If you move the pens back to the original position, they will say that the two are the same length again. On the other hand, an older child will probably think you are crazy to propose that the two pens might change length just because you moved them. The older child has learned or developed **conservation of length** while the younger child has not. If you push a young child who has not learned to conserve, you might be able to convince that child to say that the objects have the same length just to agree with you, but this does not mean that he or she understands the concept of length.

Children need experiences to help them develop conservation in other types of measurement as well.

Conservation of area involves the idea that a specific number of square units, say 20, make up the same area whether they form a four by five rectangle or are spread out on paper.

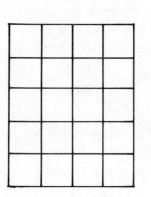

Conservation of volume means understanding that when water from a tall glass is poured into a flat pan, there will still be the same amount of water—or that a tall shape made of twelve blocks has the same volume as a short one made of twelve blocks.

Conservation of weight means being sure that a ball of clay maintains the same weight even after it has been flattened into a pancake or rolled into a long rope-like shape.

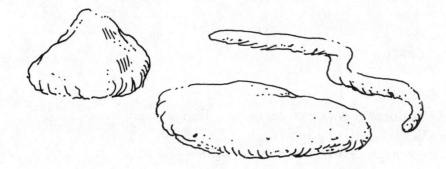

Children need many experiences in measurement, some of which are found in this chapter, before they go on to more advanced activities. No amount of rushing or telling will teach them the concepts of conservation, just lots of experiences.

This section consists of a series of experiences for each measurement topic, with only a few activities spelled out in complete detail. The sequence for each topic follows the order in which children naturally learn about measurement:

☐ first, making comparisons of two objects—which is taller? which heavier?

☐ next, ordering objects by size—from which holds the most to the least;

☐ then, measuring using common objects such as toothpicks for non-standard units;

☐ finally, measuring using standard units from both the English and Metric systems.

While it is important for chidren to become familiar with both measurement systems, it is not necessary for them to make exact conversions from one system to the other. This will not be important until high school physics or later. A sense of the size of metric units and an ability to make estimates using these units is what matters now.

A summary of metric prefixes and rough comparisons between metric and English units is below.

METRIC SYSTEM INFORMATION

Based on Tens:

The prefixes tell you what fraction or multiple of the unit you are using

kilo	1000 ×	
hecta	100 ×	
deca	10 ×	
deci	.1	(1/10 of)
centi	.01	(1/100 of)
milli	.001	(1/1000 of)

Example:

A **milli**meter is 1/1000 of a meter. That is, there are 1000 millimeters in a meter.

How many **centi**meters in a meter?

Basic Metric Units:

length	meter
area	square meter
volume/capacity	cubic meter/liter
	(1000 cubic centimeters in 1 liter)
mass/weight	gram
temperature	° (degrees) Celsius

You May Want to Remember:

1 liter is a little more than a quart
1 meter is a little more than a yard
1 kilogram is a little more than 2 pounds

$100°C$ = boiling
$37°C$ = body temperature
$30°C$ = warm day
$20°C$ = room temperature
$0°C$ = freezing

One final word as you work through the measurement activities. Always estimate first! You will be surprised at how quickly the whole family's estimation skills improve with practice. Remember, there is no rush to do every single activity, especially with young children. They will need many repetitions at each stage.

UNITS OF MEASUREMENT

		Standard or English	**Metric**
Length	How long is it? How wide is it? What is its circumference?	inches (″) (in.) feet (′) (ft.) yards (yd.) miles (mi.)	millimeters (mm) centimeters (cm) meters (m) kilometers (km)
Area	How much does it cover? How much does it take to cover it?	square feet (ft.^{2}) acres	square meters (m^2) hectares (ha)
Capacity and Volume	How much space does it fill? How much does it hold?	cubic feet (ft.^{3}) cubic inches (in.^{3}) quarts (qt.) gallons (gal.) bushels (bu.)	cubic meters (m^3) cubic centimeters (cm^3) liters (l)
Weight or Mass	How heavy is it?	ounces (oz.) pounds (lb.)	grams (g) kilograms (kg)
Temperature	How hot is it? How cold is it?	degrees Fahrenheit (°F)	degrees Celsius (°C)

1 INCH Graph paper

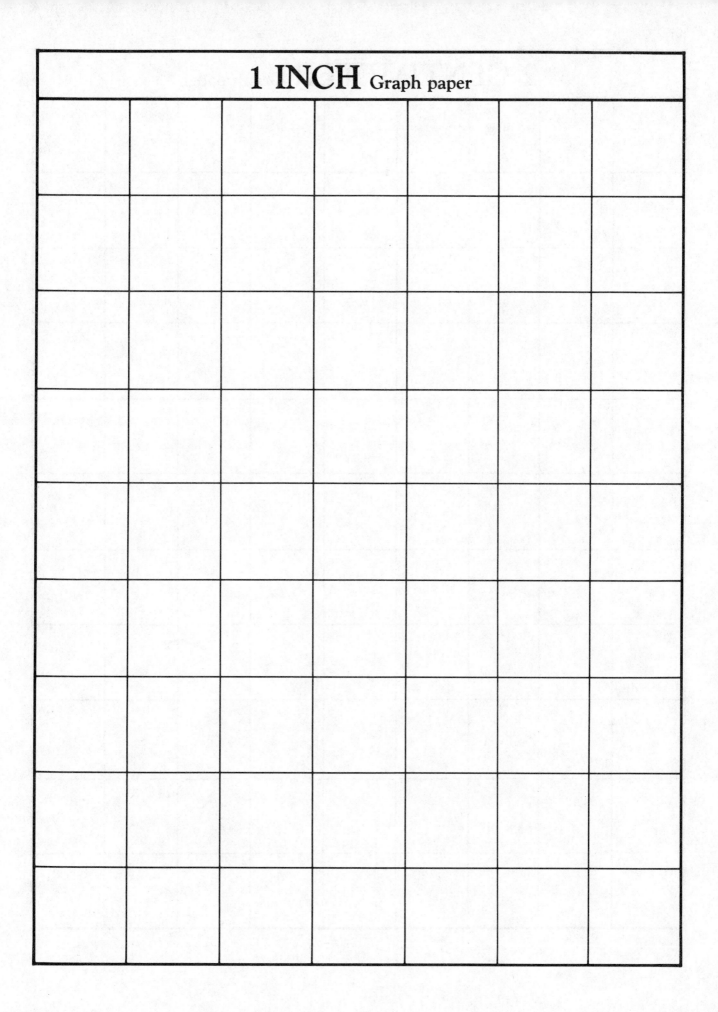

2 CENTIMETER Graph paper

1/2 INCH Graph paper

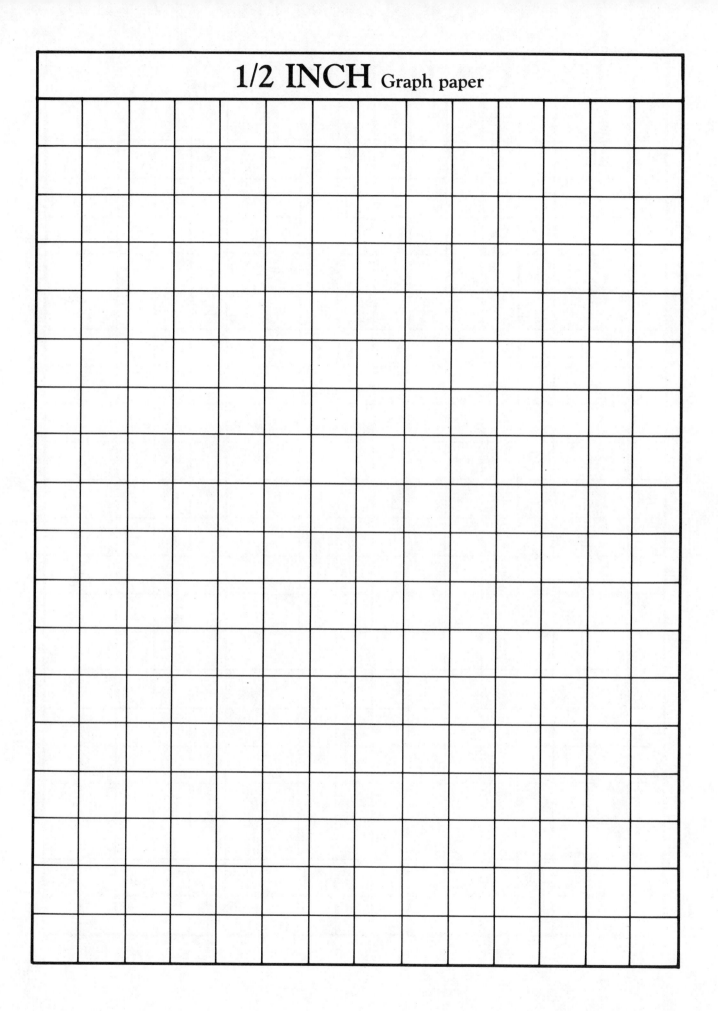

1 CENTIMETER Graph paper

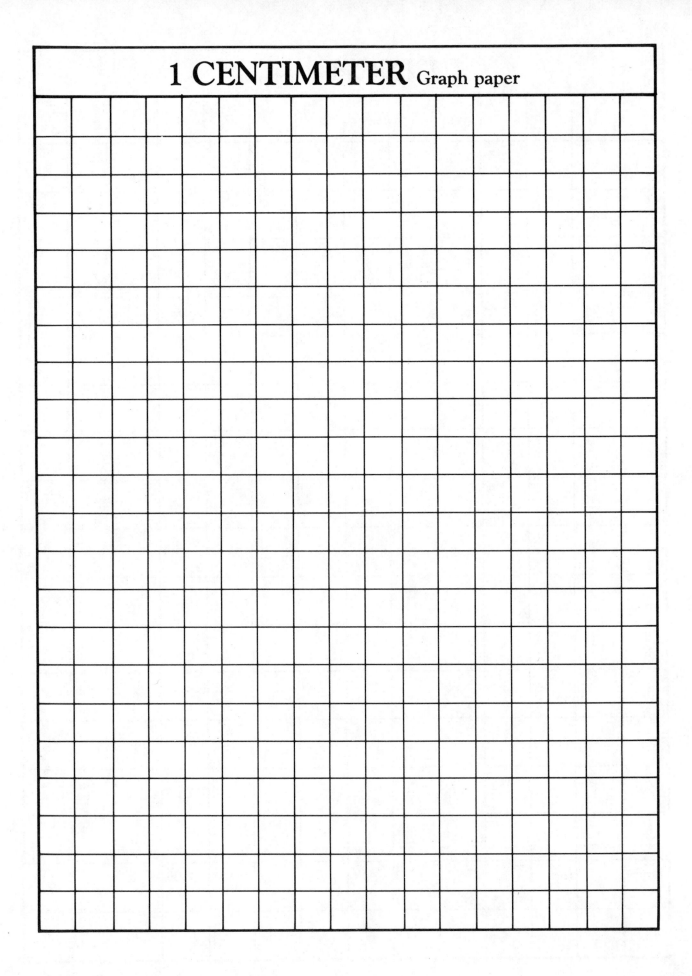

Comparing and Ordering

Why

To experience and observe comparisons between sizes of a variety of objects

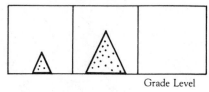

TOOLS

Toothpick, pen, or pencil
Drawing paper

How

☐ Have each person in your family compare his or her height to other objects.
  ☐ Find something taller.
  ☐ Find something shorter.
  ☐ Draw pictures to illustrate yourselves and these objects.
☐ Pick an object like a toothpick, pen, or pencil to be your "unit" of length.
  ☐ Find five objects that are longer than your "unit."
  ☐ Find five that are shorter and five that are about the same length.
  ☐ Draw a picture or make a chart of what you found out.

(UNIT)

SHORTER	ABOUT THE SAME	LONGER
PIN	PAINT BRUSH	TABLE

☐ Draw a picture of your family standing in order from tallest to shortest.
☐ Pick out five jars, cans, bottles, or books.
  ☐ Put them in order from shortest to tallest.
  ☐ Talk about whether it makes a difference if something is very round and fat.
  ☐ Does everybody agree on which things are taller?
  ☐ Do this again with other objects.

Toothpick Units

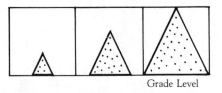

Grade Level

TOOLS

Toothpicks
Paper clips
Beans

Why

To develop understanding of the meaning of units of measurement by using non-standard units

How

☐ Get out ten toothpicks.
   ☐ Imagine the toothpicks laid out end to end on your table.
   ☐ Estimate how far they would reach. Put a pencil down where you think they will end, as in the sketch.
   ☐ Then lay the toothpicks out to see how close you come to your estimate.
   ☐ Try this activity again with ten paper clips, ten beans, and other objects.

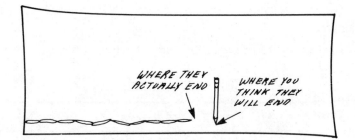

WHERE THEY ACTUALLY END

WHERE YOU THINK THEY WILL END

☐ Find the length of other objects using toothpicks.
   ☐ Draw pictures to show what you find.
   ☐ Encourage your children to estimate half units when this is appropriate.
☐ Now measure the length of objects using just one toothpick.
   ☐ Move it carefully along the length of this book, counting as you go.
   ☐ Measure some other objects this way.
   ☐ (Note: Do not begin this activity until your chidren are quite comfortable using many toothpicks.)

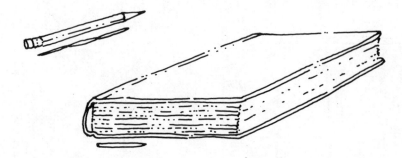

Perfect People

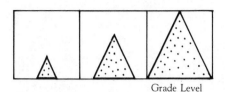

TOOLS

**String or
  adding machine tape**

Why

To explore measurements of people

How

☐ Have your family help each other cut a string or adding machine tape to exactly the height of each person. Each person should use his or her own string or tape to complete the rest of the activity.

☐ Use your own string or tape to find out if you are a **"tall rectangle,"** a **"short rectangle,"** or a **"perfect square."** To do this, have someone help hold your string or paper along your outstretched hands.

   ☐ If the string is longer than your reach, you are a tall rectangle.

   ☐ If your string is shorter than your reach, you are a short rectangle.

   ☐ If the string is about the same length as your reach, then you are a perfect square.

   ☐ Record yourself on a chart like the one below.

SHORT-RECTANGLE	PERFECT-SQUARE	TALL-RECTANGLE
MATHEW	GEORGE	LINDA
SUE		BOB

☐ Next use your paper or string to compare your height to the distance around (circumference of) your head, waist, and wrist. Write a description of what you found, like Tom's. Compare your results with others of your family or your friends.

TOM
MY HEIGHT IS — 3 TIMES THE CIRCUMFERENCE OF MY HEAD
MY HEIGHT IS — 2 TIMES THE CIRCUMFERENCE OF MY WAIST
MY HEIGHT IS — 11 TIMES THE CIRCUMFERENCE OF MY WRIST

The Magnified Inch

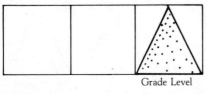

Grade Level

TOOLS

8-1/2"×14" paper or longer
  (1 piece per person)
Pen or pencil
Rulers marked in
  1/2, 1/4, 1/8 inches

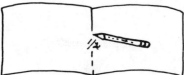

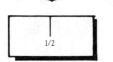

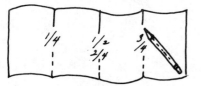

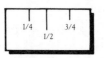

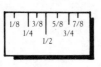

Why

To explain the lines representing fractions of inches on rulers

How

- Pretend that the length of one inch has been magnified to the length of your paper.
- Fold the paper in half end to end.
  - How many sections does your paper have?
  - Draw a line along the fold about three inches long.
  - Write 1/2 under your line to indicate it is 1/2 of the way along the paper. Compare your paper to a ruler marked in half inches.
  - Use your magnified half-inch ruler to measure objects to the nearest half inch.
- Now fold your paper in half again.
  - Draw a line about 2-1/2 inches long on each new fold.
  - The first line is 1/4 of the way along the length of your paper.
  - Write 1/4 under it.
  - The next line is 2/4 or 1/2 of the way along.
  - The third line is 3/4 of the way along.
  - Write 3/4 under this line.
  - Compare your paper with a ruler marked by quarter inches.
  - Use your magnified quarter-inch ruler to measure some objects to the nearest quarter inch.
- Now fold your paper in half again.
  - How many sections are there now?
  - Fill in the rest of the numbers.

More Ideas

- Continue this process with one more fold to indicate sixteenths. With two more folds to indicate thirty-secondths.
- Measure objects to the nearest sixteenth or thirty-secondth of an inch.
- Learn to read millimeters and centimeters on metric rulers.

Activities with Area

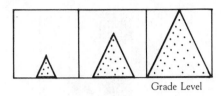

Grade Level

Why

To provide experiences in finding area of many objects using a variety of units

How

TOOLS

Circular Lids

Measurement Tool Kit

☐ Pick a circular lid.

☐ Find some lids that are smaller.

☐ Find some lids that are larger.

☐ Put your lids in order by size.

Mitten Hands

☐ Trace around your hand with your fingers together to make a shape like a mitten.

☐ Guess how many beans it will take to cover your shape. Check.

☐ Guess how many one-inch squares it will take to cover this shape. Check.

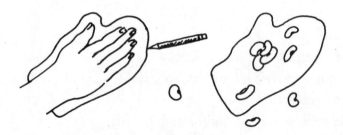

Square Inches

☐ The area of an object is the number of square units that cover it.

☐ First guess how many one-inch squares will cover each of the following objects: piece of paper, a record cover, a magazine, your favorite book, a record. Then check.

☐ Put these objects in order by area.

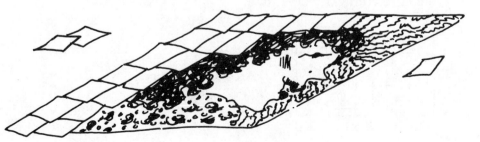

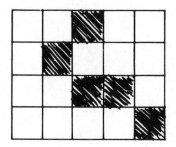

Color Design

☐ Color a design with any 36 squares on a 10×10 piece of graph paper.

☐ Make several more designs with 36 squares.

☐ Compare your designs with a friend.

☐ Remember, they all have the same area—36 square units. (For younger children, color any five squares on a 4×5 piece of graph paper.)

Square Centimeters

☐ Find five objects that are smaller in area than a square decimeter and five that are larger. A square decimeter is 10 centimeters by 10 centimeters.

☐ Set the objects onto centimeter graph paper to help you compare their areas.

☐ Do the same activity using a square foot and square inches.

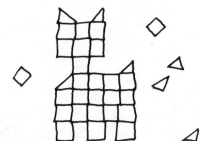

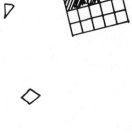

Inch Designs

☐ Use square inches ☐ and half inches ◿ to make a design.

☐ Record the area of your design.

☐ Make another design and record its area.

Area: 21 square units

Partial Squares

☐ Trace odd-shaped objects on square centimeter graph paper.

☐ Approximate the area of each by adding the number of full squares that are covered to the number of partial squares that are covered divided by two.

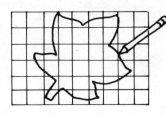

Area: approximately
23 square centimeters

Box Faces

☐ Trace all six faces of a box on square centimeter graph paper.

☐ Find the area of each face and add them to find the surface area of the box.

☐ Do this for another box.

Centimeter paper

Expanded Rectangles

☐ Draw a 3 cm × 2 cm rectangle.

☐ What happens to the area when you double the length?

☐ Double the width?

☐ Double both the length and the width?

☐ Answer these questions for a 5×4 rectangle and a 6×6 square.

Square Meter

☐ Make a square meter. Cover it with decorated square decimeters.

☐ Make a square yard.
Cover it with decorated square feet.

Comparing Squares

☐ Figure out how many square centimeters in a square decimeter, in a square meter.

☐ Figure out how many square inches in a square foot, in a square yard.

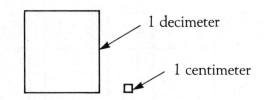

1 decimeter

1 centimeter

Comparing Perimeters

☐ On centimeter graph paper, draw a shape with a perimeter of fourteen centimeters.

☐ Find its area.

☐ Draw a different shape with the same perimeter.

☐ Does it have the same area?

☐ Can you find one with a different area?

☐ What short of shapes with a perimeter of fourteen have the largest area?

☐ Do this activity for a perimeter of sixteen.

Computing Area

☐ The area of an object is the number of square units that cover it.

☐ How could you find the area of a rectangle six inches long by four inches wide if you had no graph paper or square inches to use?

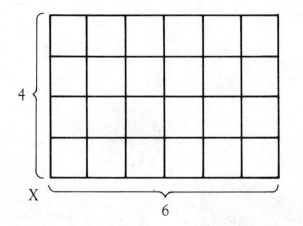

4 {

X

6

Capacity and Volume

Why

To develop understanding of volume relationships

▶ *Children should have many experiences with volume, comparing, measuring with non-standard units, and measuring with standard units. Opportunities for this kind of activity are sometimes limited at school, so it is important to provide them at home. Sometimes it is fun to put a little food coloring into the water, and do the activities out in the back yard, in the bathtub, or at the kitchen sink.* ◀

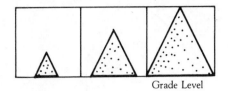

Grade Level

TOOLS

Containers of all kinds and sizes

How

Half Cup

☐ Help your child fill a small container with water. Then have him or her pour the water into a different-shaped container, and observe what happens.

☐ Pour the water into still another container and see what it looks like.

☐ Next pour the water into a measuring cup, up to the **half-cup mark** if there is enough.

▶ *Talk about how the water looked in the various containers, without being formal about measurement. Experiences of this kind develop intuitive understanding of volume, a prerequisite for formal understanding.* ◀

☐ Repeat this activity using sand, rice, or beans.

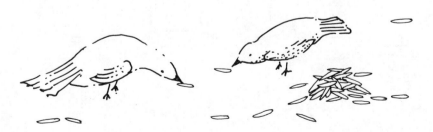

Full Glass

☐ Find a glass of medium height and have your child fill it with water.

☐ Pour the water into another shorter but fatter container.

☐ Notice how high the water goes in this container. Pour the water back into the glass.

☐ Find a container that is taller than the glass. Guess how high the water will go in this container, and then pour it in. Was your guess too high or too low?

☐ Pour the glass of water into other different-shaped containers, guessing first each time how high it will go.

Five Containers

☐ Label five containers A, B, C, D, E.

☐ Choose one of the containers, and fill it with small beans.

☐ Choose another of the containers, and decide whether the new container holds more, the same, or less than the first one.

☐ Then pour the beans from your container into the new one. Did you guess right?

☐ Record your findings. Compare how much your container holds with how much the others hold.

☐ Arrange the containers in order, from smallest to largest.

Holds less   Holds more

Container Units

- ☐ Have your child choose the smallest of your containers to be used as a "unit" of capacity.
- ☐ Using water, beans, rice, or sand, measure how many of your units it takes to fill the different containers.
- ☐ Record your results. You may want to put a piece of tape onto each container and write the number of "units" it will hold.

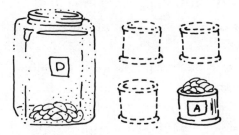

Ordering Containers

- ☐ Arrange three (or five) containers in order by **estimated** capacity—from "holds the most" to "holds the least."
- ☐ Use a small container and rice or beans to help you check the arrangement.

Measuring Cup

- ☐ Measure how much different containers hold by filling them with water and then pouring the water into a standard measuring cup.
- ☐ Do the same using a liter container marked in milliliters or cubic centimeters.

► *Note that one milliliter of water fills one cubic centimeter of volume.* ◄

Rock Volume

☐ Fill a plastic container about half full with water. Mark the water level with masking tape.

☐ Then carefully drop in a small rock. What happens to the water level?

☐ Take the rock out and put in another heavy object. What happens this time?

Margarine Volume

☐ Fill a pint or quart measure to the one-cup mark with clean water.

☐ Carefully push a quarter-pound cube of margarine completely under the water.

☐ How much does the water level change? What does this tell us about the cube of margarine? Can you figure out its volume?

☐ Do the same activity with a half-cube of margarine. What is the volume of the half-cube?

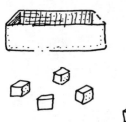

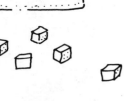

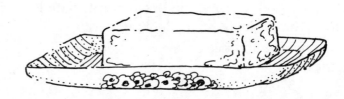

Sugar Boxes

☐ Label some small boxes A, B, C, D, and E.

☐ Use sugar cubes or other cubes to fill the boxes.

☐ Record your results, or label the boxes.

Eight-Cube Solid

☐ Use eight cubes to build a rectangular solid (a solid figure that looks like a box, with no indentations).

☐ Build as many different solids with eight cubes as you can.

☐ Note that all of these solids have the same volume—eight cubic units.

☐ Do the same activity with twelve or sixteen cubes.

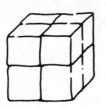

More Cubic Solids

☐ How many cubes does it take to build a 1×2×3 rectangular solid?

☐ How many cubes does it take to build a 2×4×6 rectangular solid?

☐ The number of cubes it takes to build it is the **volume** of the solid.

☐ Find the volumes of a 1×1×1, a 2×2×2, and a 3×3×3 cube.

☐ Make a table showing your results.

LENGTH OF ONE SIDE	NUMBER OF CUBES
1	1
2	8
3	?
4	?
5	?

Circumferences

- Use a string to compare the **circumference** (distance around) to the **height** of different jars, bottles, and other round containers.
- First put the string around the container and hold it with your finger to mark the circumference.
- Then straighten it out to compare with the height. Are you surprised at any of your results? What do you think is happening?
- A good comparison to make to help you think about this problem is the circumference of a tennis can to its height.

Perimeter Variations

- Arrange six square tiles (or paper squares) so that each square touches at least one other along a full side.
- Count the outside edges to find the perimeter of your shape.
- Make some more six-square shapes and find their perimeters.
- Are all the perimeters the same?
- Do the same with fifteen tiles (or squares).
- Are all the perimeters the same?
- What sort of shapes have the longest perimeters?
- What sort have the shortest perimeters?

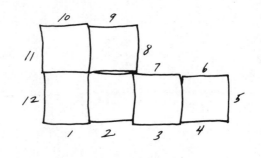

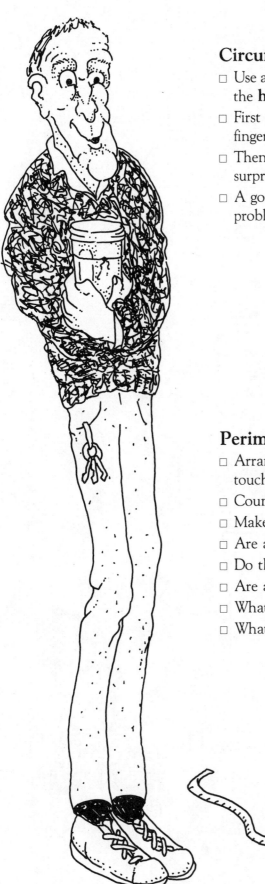

Lid Ratios

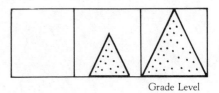

Why

To develop understanding of the concept of **π**, or the relationship between the circumference of a circle and its diameter

How

☐ Pick out a lid to measure.

☐ Cut a ribbon or string that measures around the lid exactly (the circumference).

☐ Cut a ribbon or string that measures across the center of the lid (the diameter).

☐ Tape your circumference and diameter ribbons onto a large sheet of paper.

☐ Cut ribbons to measure the circumferences and diameters of several more lids.

☐ Tape these ribbons on your record paper.

Grade Level

TOOLS

Circular lids
Ribbon or string
Scissors
Large sheet of paper
Pen or pencil

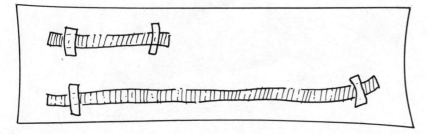

☐ Study the circumference and diameter ribbons for the lids.

☐ About how many times longer is the circumference ribbon than the diameter ribbon?

☐ How many diameter ribbons would fit along the circumference ribbon?

▶ *The formula for calculating the circumference is*

$$Circumference = π \times Diameter \qquad C = πd$$
or   *or*
$$Circumference = 2 \times π \times Radius. \qquad C = 2πr$$

The pi ratio of about 3.1416 is the same for every circle. ◀

More Ideas

Measure the exact lengths of the circumference and diameter ribbons for each lid. Use a calculator to find the ratio $\dfrac{\text{circumference}}{\text{diameter}}$ for each lid. Is there a pattern?

Making Balance Scales

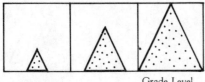

Grade Level

TOOLS

See each suggested project

Why

To experiment with comparing weights

▶ You may be able to buy balance scales from a hardware or variety store, or at a garage sale. You may also be able to borrow them from the nearest school. Professional scales are a great help for serious projects.

There is a great deal of benefit, however, in helping children experiment with making their own balance scales. Here are some ideas to start with, but you and your family will have many more.

Some of these scales may not work very well, so it may take really heavy things to show a difference between one object and another. Others will work very well, with fine balancing. Part of the difference depends on how carefully they are made. ◀

How

□ Use a round log, such as firelace wood, and balance a long board on top of it. When two people stand an equal distance from the center, which end goes down? How close will the larger person have to move before the smaller person's end goes down?

☐ Use a small round object, such as a pencil, and balance a ruler on top of it. Put a small paper cup carefully on each end of the ruler, and re-balance. Put beans or other small objects into the cups. Experiment with various objects. How many paper clips balance a marble?

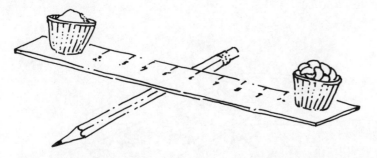

☐ Use a coat-hanger, with two paper clips and two light-weight paper or plastic cups. Make a balance that looks like this, and hang it from a hook or the top of a door or window. Try putting various objects into the two cups to do balancing experiments.

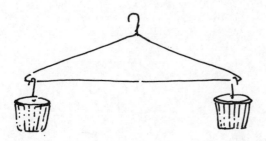

☐ Use an *old* ruler (that you don't care about) or a stick about the size of a ruler. Drill three holes in it:

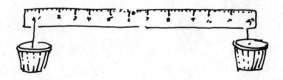

Fasten paper clips and cups to each end, a string to the center, and hang the structure from a nail or the top of a door. Add very small weights (beans or bits of clay) until the scale balances.

☐ Cut a clothes-hanger with a wire cutter.

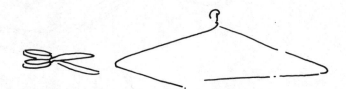

Remove the top of an egg carton, saving the partitioned portion. Push the straight part of the hanger through the center of the egg carton. Hang the hanger from a door-top, or have someone hold it. Add very small weights until the scale balances.

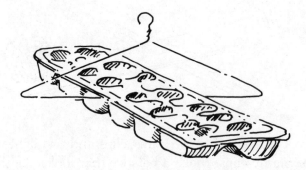

☐ Now think up some balance scale ideas of your own.

Weight Activities

Why

To see and feel graphic representation of the terms **heavier than,** **lighter than,** and **weighs about the same**

► *Borrow or make a balance scale to use for these activities.* ◄

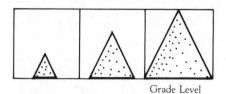

TOOLS

Balance scale (see pages 98-100)
Paper clips
Various objects
Small blocks

How

Which is Heavier?

☐ Pick up two objects. Put one in each hand and try to decide which is heavier.

☐ Check by putting the objects on opposite sides of a balance.

☐ Do this activity again with many different objects.

On Balance

☐ Find two objects that you think weigh the same

☐ Put them on the balance scale to check.

Lightest to Heaviest

☐ Use a balance to put in order three objects (or five objects) from lightest to heaviest.

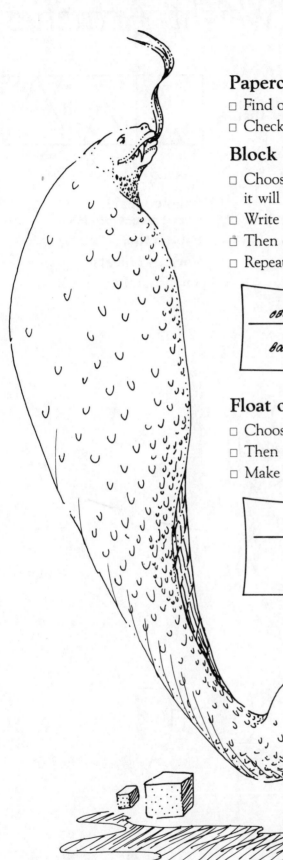

Paperclips and Blocks

☐ Find out how many paperclips balance one block.

☐ Check to see how many paperclips balance a second block.

Block Weights

☐ Choose an object and guess how many blocks or paperclips it will weigh.

☐ Write down your guess.

☐ Then check, using a balance.

☐ Repeat with other objects. Record your results on a chart.

OBJECT	GUESS	ACTUAL WEIGHT
BOOK	15 BLOCKS	16 BLOCKS AND 2 PAPER CLIPS

Float or Sink

☐ Choose an object and guess whether it will float or sink.

☐ Then test in a container of water. Repeat with other objects.

☐ Make a chart of your results.

FLOAT	SINK
WOOD BLOCK SPONGE	ROCK PAPER CLIP

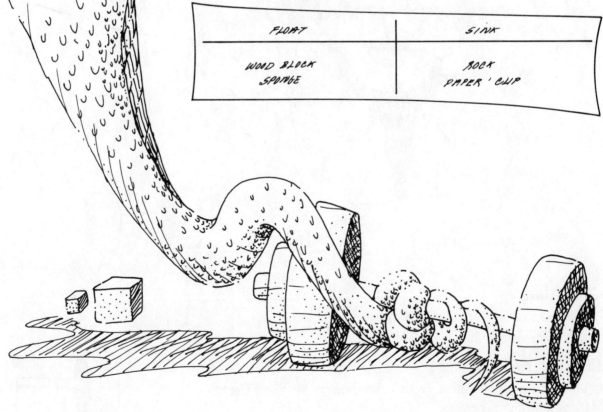

Temperature

Why

To gain intuitive understanding of the relationship between hot and cold temperatures and the numbers on thermometers

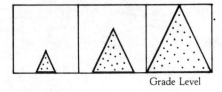
Grade Level

How

Sandpaper

- ☐ Firmly rub a piece of sandpaper rapidly back and forth on a piece of wood
- ☐ Stop and feel the surfaces of the wood and sandpaper.

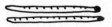

TOOLS

Sandpaper
Piece of wood

Comparing Thermometers

- ☐ Put a Celsius thermometer and a Fahrenheit thermometer next to each other
- ☐ Compare the temperature readings at different times.

Celsius and Fahrenheit thermometers

White and Black

- ☐ Put a piece of white paper and a piece of black paper in the sun.
- ☐ After five minutes, which feels warmer?
- ☐ Check again after half an hour.
- ☐ Repeat this activity with a thermometer under each paper.
- ☐ Read the temperatures after five minutes and 30 minutes.

White and black paper
Thermometers

Graphing Temperatures

- ☐ Make a graph of the daily high temperatures in your town for a week or for a month.

Daily paper

Ice Cubes

- ☐ Put ice cubes in water, then read the temperature of the water.
- ☐ Add rock salt to ice, and read the temperature of the ice. What does the salt do to the temperature of the ice?

Ice cubes
Thermometer
Rock salt

Chapter V

NUMBERS and OPERATIONS

TOOL KIT

Scratch paper

Construction paper of 5 to 8 different colors

Butcher paper or other long strips of paper

Pencils and pens

Scissors

Paste or tape

Dice

Play or real money (dimes and pennies)

A deck of cards

Graph paper (see page 79-82)

Tile squares (paper squares may be used)

Beans (Red beans, Lima beans, Black-eyed peas, or buttons or other objects to represent the beans)

NUMBERS AND OPERATIONS

Jamie goes to the hardware store and buys ten nails for a quarter. Later, she decides not to hammer, and she sells the nails to her friend Tim for two quarters. The next day, Jamie finds a project (putting up five posters) that requires nails. Tim, good friend that he is, sells the nails back to Jamie for three quarters. Before Jamie uses the nails, her friend Tommy comes over with a staple gun. They use staples instead of nails to secure the posters in their clubhouse. Fortunately for Jamie, the neighborhood carpenter, Sean, comes by to see if there are any nails left. Jamie sells Sean the ten nails for four quarters. What was the outcome of Jamie's transactions? Did she lose money, gain money, or break even?

Even though children can do all of the operations involved in this problem, they often don't have enough experience in reasoning through complex situations like this one. Instruction in numbers and operations should provide youngsters with the ability to use arithmetic to solve problems.

Simply teaching the "rules" for addition, subtraction, multiplication, and division, is not enough.

The view that mathematics is arithmetic and that arithmetic is a group of rules may have been a universally held belief at one time, but good math programs now include a much broader curriculum, with teaching for understanding instead of following rules. Even so, teachers of elementary school children often feel pressured to cover skills areas quickly, without enough time to allow children to experience arithmetic sufficiently. As a result, we have a nation of children who can compute the answers to arithmetic problems but are not capable of applying that computing power effectively.

Arithmetic should not be taught as a series of rules because:

Rules are easy to forget, especially if they are learned without an understanding of their applications. Children often use a combination of their own logic with a half-remembered rule: $1/2 + 1/4 = 2/6$; 10% of $100.00 is $1.00; $42 - 28 = 26$; or $\square - 4 = 5$ means 1 goes into the $\square$. Each of the above errors has a logical basis and is the result of using "rules" without understanding their meaning.

Rules minimize thinking. Doing arithmetic solely by the rules enables a child to manipulate numbers symbolically without thinking about the values of the numbers, or what action is really happening. Ask the following question: If you cut ten feet of cloth into two-inch strips, how many strips will you have? A typical response is five.

Rules prevent visualizing interrelationships. Learning arithmetic by the rules makes it difficult to visualize interrelationships in mathematics; adding whole numbers is seen as something different from adding fractions. Fractions are seen as unrelated to decimals except for certain pages in the textbook where one has to change fractions to decimals and decimals to fractions according to a rule. Similarly, 2/4 can be reduced to 1/2 but often children are uncertain which is larger, 1/2 or 2/4, not realizing that 1/2 = 2/4.

Rules do not work well for problem-solving. Children who only do arithmetic by the rules are being cheated of the chance to put their skills to work in problem-solving situations.

Such a soapbox!! Clearly, FAMILY MATH does not advocate rote memorization of rules as the principal method of learning numbers and operations. What should you do at home to increase the likelihood that your child will have an understanding of the arithmetic she or he has learned in school? First of all, **talk** with your child about arithmetic. What does that "1" mean in 16? Can you draw a picture about this problem? Which number is larger? What do 1/2 and 2/4 look like?

Have outrageous story contests. Start with "I'm going to make up a story problem that uses 3×4 to get the answer. See if you can make up one more outrageous than mine." Here's a 3×4 story created by a learning disabled sixth-grader:

> It was well after midnight when three trolls (Zapp, Evarista, and Uglo) crept out of their cottage. Tonight was the night they would recover the stolen treasures. The treasures were buried in a sewer far below the Mayor's house. Each of the trolls made four trips into the sewer. Every time one came back up, she had a bag of loot. How many bags of loot did they have altogether?

Have discussions about numbers. "Tell me twelve things about the number **6**." "Is zero a number?" "Tell me, what happens if we divide one by one?" "If I were from Ethiopia and had never seen the number 3/4 but could speak English, what ten things could you tell me to help me understand all there is to know about 3/4?" "Look at this picture (or group of objects)—tell me what arithmetic problem it could represent."

When you do the activities in this chapter, remember that the children often need concrete references for the symbols they manipulate. For "Double Digit," (page 111) for example, you may want your youngster to take a popsicle stick for each ten and a bean for each unit as they count to 100, instead of using only paper and pencil.

The bottom line in learning multiplication, subtraction, and addition facts is memorization. Even memorizing can be fun, and the process can be made much easier if a child sees patterns in the numbers and has an understanding of the operations.

We think arithmetic is full of wondrous and elegant notions. Besides learning to compute, children benefit from enjoying the computations, appreciating the beauty and structure of numbers. Try to keep or regain the natural exuberance most children have about numbers. Before these games become drudgery, or a discussion becomes a quiz session, relax and stop for a while. The activities should be used to enrich your relationship with your child, and your child's relationship with mathematics, not to create more stress.

Learning the Basic Facts

Why

To provide alternative methods of learning and practicing the basic arithmetic facts of addition, subtraction, multiplication, and division

> ► *Parents, teachers, and students themselves will usually identify "learning the basic facts" as one of the major steps in mathematics. Here are some ways to help people with this step. Try them all, to find out which has most appeal and success for your family.* ◄

Practice with just one number fact on a given day. For example, several times in the morning, repeat together "seven times eight is 56." Then suggest to your child that the same number sentence be repeated whenever possible all day long. Usually by the end of the day, that particular fact is well learned. You'll be surprised at how quickly the days will go by, and the facts all memorized.

Repeat the "fact of the day" in as many voices as possible —shout it, sing it, say it in a rumbly voice, use a piping voice, go up the scale or down the scale, and so on—use plenty of imagination. How would a fish say $5 \times 9 = 45$?!

Look at the addition or multiplication table and cross out all the facts that are already known, such as the "one times" and the "two times." Then cross out one example of those that are shown twice. The problem 6×7 is the same as 7×6! If you look at what is left to be learned, it doesn't look quite so hard as thinking that the entire table has to be memorized.

Work from answers back to the problems. There is only one of the basic multiplication facts that has an answer in the 80's ($9 \times 9 = 81$), and only one in the 70's ($9 \times 8 = 72$). How many in the 60's? ($8 \times 8 = 64$, $9 \times 7 = 63$) How many in the fifties? (Only two— $7 \times 8 = 56$, and $9 \times 6 = 54$.) Now *that* is not many to learn is it? And usually these particular facts are the most difficult. See also the "One Hundred Cards" activity on page 44.

Concentrate on learning the "square" numbers first. Those are the factor times itself, or the addend plus itself. Once memorized, these will give a basis to which we can return when we forget some of the other facts. For example, if 7×8 is difficult, remember that $7 \times 7 = 49$, and one more 7 makes it 56.

$1 + 1 = 2$	$1 \times 1 = 1$
$2 + 2 = 4$	$2 \times 2 = 4$
$3 + 3 = 6$	$3 \times 3 = 9$
$4 + 4 = 8$	$4 \times 4 = 16$
$5 + 5 = 10$	$5 \times 5 = 25$
$6 + 6 = 12$	$6 \times 6 = 36$
$7 + 7 = 14$	$7 \times 7 = 49$
$8 + 8 = 16$	$8 \times 8 = 64$
$9 + 9 = 18$	$9 \times 9 = 81$

Practice "fact families," to understand the relationship between addition and subtraction, or between multiplication and division. For example, $6 \times 7 = 42$, $7 \times 6 = 42$, $42 \div 6 = 7$ and $42 \div 7 = 6$ make up a "fact family," since all of the problems show the same number relationship, stated in different ways. This is especially valuable when the addition or multiplication facts have been learned, and the subtraction or division facts are proving difficult.

Be sure your child really understands the meaning of the numbers and of the operation. Use objects to demonstrate the problem if possible, or diagrams when the numbers become too large.

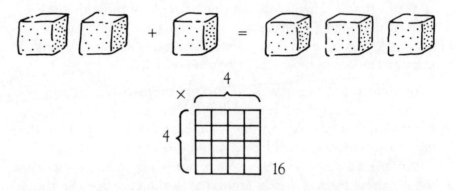

Use flash cards with caution. If a child does not really know the answers anyhow, a pure guess may not contribute to learning.

Study the patterns in addition and multiplication charts. See the chapter on Number Charts for additional activities.

Make a little chart of the arithmetic vocabulary, such as:
Addend + Addend = Sum
Minuend − Subtrahend = Remainder
Factor × Factor = Product
Dividend ÷ Divisor = Quotient & Remainder

Double Digit

Why

To practice place value and estimation skills

> ▶ *Both skill and chance play important roles in this game. The dice rolls make it difficult to use a consistent winning strategy. However, an intuitive understanding of probability, or what* usually *happens, will allow children to find a strategy that will be successful more often than not. Development of estimation skills will increase a child's chances for success in other areas of mathematics.* ◄

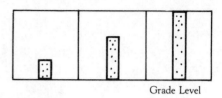

Grade Level

TOOLS

Pencil
Paper for scoresheets
Dice

*A game for
2–6 players*

How

☐ You will need one die for the group and a scoresheet for each person in your family, like this one:

TENS	ONES
1.	
2.	
3.	
4.	
5.	
6.	
7.	

☐ Each person takes a turn rolling the die.
☐ The number may be written in either the tens' column or the ones' column of the scoresheet.
  ☐ When a number is entered in the tens' column, "0" is written next to it in the ones' column. Thus, 4 written in the tens' column counts as 40.
☐ After each player has rolled the die **seven** times, the players add up their numbers.
☐ The players who are left in the game compare their totals.
☐ The player who is closest to 100 without going over is the winner.

More Ideas

☐ At the end of the game, talk about what the best total could have been with those seven rolls.
☐ See also the game Dollar Digit, for younger children.

Dollar Digit

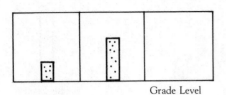

TOOLS

Scratch paper

**Play or real money—
dimes and pennies**

Dice

*A game for
2-4 players*

Grade Level

Why

To practice place value skills using money

▶ *This game is meant for younger children to help them learn about the relative values of numbers in columns, building on their understanding of money.* ◀

How

☐ Each person will need a playing board made from scratch paper, like this:

DIMES	PENNIES
1.	
2.	
3	
4.	
5.	
6.	
7.	

☐ The dimes and pennies should be in the center of the table within reach of all players.

☐ Each person takes a turn rolling one die.

☐ All of the players use the same number rolled on each turn.

☐ Each player takes as many pennies **or** dimes as the number rolled on the die, and puts them in the appropriate column— pennies in the pennies' column or dimes in the dimes' column. A player **may not take both pennies and dimes on the same turn.**

☐ Whenever a player has ten or more pennies, she or he **must** trade ten of the pennies for a dime. The dime is put with any others in the dimes' column.

☐ The object of the game is to get as close as possible to $1.00. Totals may go over $1.00.

☐ As soon as each player has had seven turns, the players look to see who is closest to $1.00, or who is the winner.

More Ideas

☐ Make a rule that if a player goes over a dollar, the player is out of the game.

☐ Use pennies and nickels to reach 25¢ for very young children.

☐ Base 10 blocks may be used if they are available.

☐ Make a written record of each turn and its money amounts.

Reverse Double Digit

Why

To practice estimation and subtraction with and without regrouping (borrowing)

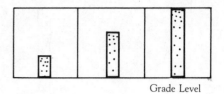

Grade Level

TOOLS

Paper
Pencil
Dice

A game for 2–6 players

▶ *This activity helps develop an intuitive understanding of probability through observation of the number of times any numeral appears and is likely to appear again. For example, if a "4" has been rolled three times, is it likely that a "4" will show up on the next turn?* ◀

How

☐ The goal is to come as close to zero as possible, without going below zero. A player is out who goes below zero.

☐ A game consists of seven turns for each player. Keep a tally count of the turns taken.

☐ Before the game begins, players write the numeral 100 at the top of their record sheets.

☐ Players then take turns rolling one die.

☐ Each player may choose to record the number he or she rolls on a turn as the number itself, or as ten times the number. For example, when 5 is rolled, it may be recorded as 5 or as 50.

☐ After each number is recorded, it is subtracted from 100, or from the remainder left from the previous turn.

☐ The game continues until each player has had seven turns or cannot subtract and is out.

☐ The person closest to zero after seven turns is the winner.

More Ideas

☐ Start with ten dimes. Remove dimes or pennies according to each roll, making change as necessary. Here a 4 can count as four pennies or four dimes. If a player does not have enough money to take away, he or she is out. The person with least money left after seven turns wins.

How Close Can You Get?

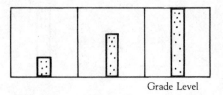

TOOLS

Deck of cards
Paper and pencil

A game for
2–5 players

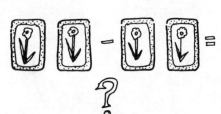

Round 2

Why

To practice subtraction, estimation, and mental arithmetic

How

☐ Remove the tens and picture cards from a deck of cards, leaving ace through nine for the game. Shuffle the cards.

☐ Deal each player four cards, face down.

☐ Turn up two more cards. The first card goes in the tens' place and the second in the ones' place to form the **target number.** For example, a six then an ace makes **61.**

Round 1

☐ Now the players turn up their four cards and arrange them into two 2-digit numbers, so that when they **subtract** their two numbers the result will be as close to the target number as possible.

☐ To find the score, each player finds how close he or she is to the **target number** by subtracting her or his result from the **target number** or vice versa, depending on which number is larger.

☐ For example, if the **target number** is 61 and a player has an A, 5, 3, 9, the best she could do is 95−31=64. Her score would be 64−61=3 for that round.

Note that you can go **over** or **under** the **target number** when you subtract. A 64 has a difference of three from 61, and a 58 also has a difference of three from 61, so either 58 or 64 would give a score of three for the target number 61.

☐ For the next round, turn up two new cards from the deck to form the next target number. Players can choose to use their same four cards and proceed as above, or deal out new cards.

☐ Add Round 2 score to Round 1 score.

☐ Play for five rounds. The player with the **lowest** total score wins.

More Ideas

☐ Play and work toward a low group score.

☐ Allow trading cards among players.

☐ Try a 3-digit **target number** and six cards for each player.

Number Line Rectangles

Why

To develop understanding of the qualities of numbers such as factors, products, prime numbers, composite numbers, and square numbers

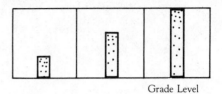
Grade Level

TOOLS

Paper or tile squares
Graph paper
Scissors
Paste or tape
Construction paper or butcher paper

► *This understanding of numbers is essential for successful work with fractions, ratio and proportion, geometry, and other parts of advanced mathematics.* ◄

How

□ Using the paper or tile squares, make rectangles of different sizes.

► *Help your children be careful about really making a rectangle. Is this a rectangle?*

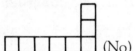 (No) 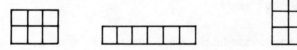 (No) ... (No)

A rectangle is a four-sided figure with all of its angles right angles. There are no missing corners. ◄

☐ How many squares are there in each rectangle? How many squares are there in the smallest rectangle?

▶ *The smallest rectangle has an area of just one square. Many people think that squares are not rectangles, but in truth, a **square is a special kind of rectangle**.* ◀

☐ What is the next rectangle size?

▶ *The next rectangle has an area of **two** squares. The "two" rectangle can be drawn this way* ⊟ *or this way* ⊡ . *These rectangles are the same size and shape. Mathematicians say they are **congruent**, so we only count one of them.* ◀

☐ What is the largest rectangle you could make if you put all your squares together?

▶ *How do you decide how large a rectangle is? One way is to count the squares that make that rectangle. Another way which we hope some children will discover for themselves is to look at the length and width and multiply the two numbers together.* ◀

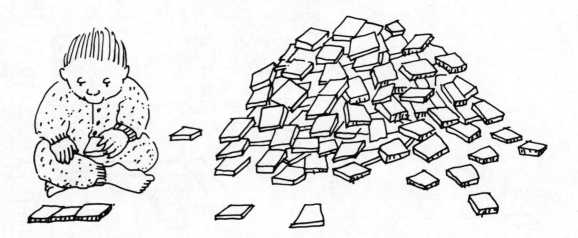

□ Now you are ready to begin using the graph paper to build a picture of all of the numbers.

□ First make a long strip of butcher paper or construction paper. Write along the top edge the numbers 1 through 25, as shown in the illustration.

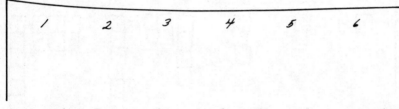

□ Start with 1. Cut a single square from the graph paper, and tape or glue it under the numeral 1.

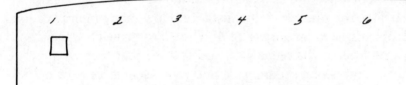

□ Continue with 2. Cut a rectangle with area of two squares, and tape or glue it under the numeral 2.

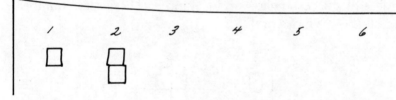

□ Cut a rectangle with area of three and put it under the numeral 3.

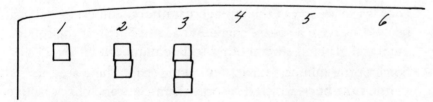

□ How many rectangles are there with an area of four? You should be able to find two—don't forget about the square.

□ If you like, label each rectangle with the number of squares it has.

☐ Continue cutting out the rectangles for each number until you have done at least 15 or 20 of the numbers, then stop and look at what you have.

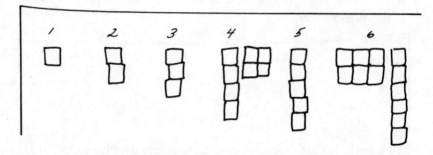

☐ We have been talking about rectangles and squares and their area, but these are also pictures of multiplication problems. Look at the number 4, for instance. How many multiplication problems have an answer of 4? There are two—1×4 and 2×2. If you look at the rectangles, you will see that there is a rectangle that is one square wide and four squares long, or a 1×4 rectangle. There is also a rectangle that is two squares wide and two squares long, or a 2×2 rectangle.

☐ You can also see that all of the **factors** of 4 are the lengths or widths of the rectangles—or 1, 2, and 4. These same numbers are all the **divisors** of 4.

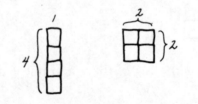

☐ Four is the second of the **square numbers.** You can see that it really does have a square shape. What is the first of the square numbers? (1) Find several other square numbers. (9, 16, 25...)

☐ Some of the numbers have only a long strip. These are the **prime numbers,** which have only themselves and one as factors.

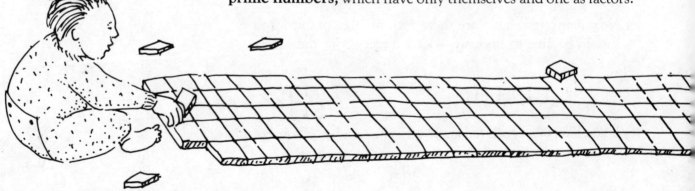

More Ideas

- You can make your family's number line of rectangles go as far as you like (or as far as you have space!). You may want to make a tiny version drawn on smaller graph paper.

- Study the numbers to find other patterns. For example, where are the numbers that have a strip three squares wide? How many of the numbers less than 25 have more than four factors? What are the common factors of 25 and 30? What is the first common multiple of 7 and 6 (that is, the first number that can be divided by both 7 and 6)?

Making a Fraction Kit

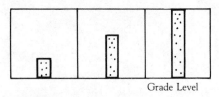

TOOLS

Pencil

Scissors

Strips of 3″ × 18″ construction paper

For Kit I you need 4 strips of different colors

For Kit II you need Kit I plus 3 more strips of different colors

Why

To see and understand the relative values of fractions by making physical representations

▸ *When young children are learning simple arithmetic, it is essential that they have many experiences with concrete materials, such as blocks, before they can truly understand the difference between three* ▱▱▱ *and five* ▱ ▱ ▱ ▱ ▱ *. In the same way, making a physical model of fractions provides reinforcement for understanding the relative values of fractions.* ◂

How

To Make Kit I

☐ Take 5 strips of different colors. With your children, compare the strips to be sure they are all the same length. Talk about the fact that the strips each represent "1 WHOLE" and that you will be cutting some into fractional parts.

☐ Label one strip "1 WHOLE." (Note: It is often convenient to use a black strip for your whole.)

☐ Take another strip and fold it carefully in half.
  ☐ Fold by first lining up the edges of the strip and then creasing the fold.

How many sections will you have when you open your folded strip?

Open it and count.

☐ Label each part 1/2 and cut on the fold line.

☐ Take another strip and fold carefully in half two times.

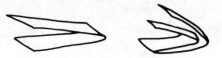

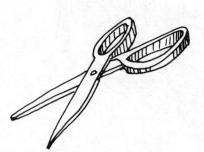

Guess how many sections you will have when you open it.

Count the sections.

☐ Label each part 1/4 and cut them apart.

☐ Take another strip. This time fold it in half **three** times.

Again, be very careful to fold accurately.

How many sections will there be this time?

Count to check.

☐ Label each part 1/8 and cut them apart.

☐ Continue with the last strip. Fold **very** carefully **four** times.
This time you will get one-sixteenth (1/16) for each section.

☐ Put your name on each part of your kit.

☐ Keep the pieces in a large envelope or a shoe box.

☐ This completes Kit I. Primary students should work with Kit
I for some time before making Kit II.

To Make Kit II:

☐ Kit II consists of Kit I plus the pieces made from three more 18-inch strips.

☐ Make Kit I.

☐ Take the next strip, measure and mark it with a pencil at 6″ and 12″ along the edge. Fold on these lines.

  ☐ You will have three sections.

  ☐ Label each 1/3 and cut them apart.

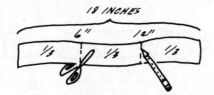

☐ Take the next strip. Make thirds and then fold each third in half.

  ☐ How many sections do you have?

  ☐ Label each section 1/6 and cut them apart.

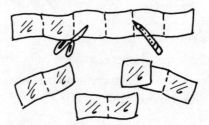

☐ Take the last strip. Make sixths and then fold each sixth in half. You will have twelve sections this time.

  ☐ Label each section 1/12 and cut them apart.

☐ Put your name on each part of your kit.

☐ Use your fraction kit to compare the sizes of different fractions and for Fraction Cover Up and other fraction activities.

More Ideas

Equivalent Fractions are easily shown with these kits. For example, 1 WHOLE is the same as 2/2, 3/3, 4/4, etc. Explore with your children some other equivalent fractions, using your strips to check. Keep a record like this:

½ IS THE SAME AS □/□ OR □/□ OR □/□

⅔ IS THE SAME AS □/□ OR □/□ OR □/□

4/16 IS THE SAME AS □/□ OR □/□

Fraction Kit Games

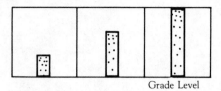

TOOLS

A Fraction Kit for each player (see pages 120-123)

One die labeled:
  1/2, 1/4, 1/8, 2/8,
  1/16, 2/16
  for Kit I or

One die labeled:
  1/3, 1/4, 1/6, 1/8,
  1/12, 1/16
  for Kit II

*Games for
2–6 players*

Why

To practice using fractional parts of a whole, recognizing relative sizes and **equivalent fractions**

> ▶ *Before your children can learn to add, subtract, multiply, or divide fractions, they must understand the relationship between different kinds of fractions.*
>
> *For example, in order to add 1/6+2/3, it is necessary to realize that 2/3 is the same as or **equivalent** to 4/6. 1/6 added to 2/3 may not make sense, but 1/6 added to 4/6 is 5/6. Changing the thirds to sixths requires finding a **common denominator,** or a fractional part that is part of both sixths and thirds.* ◄

How

Fraction Cover Up

☐ Start with your "1 WHOLE" strip in front of you.

☐ Take turns rolling the die.

☐ Take the fraction you roll and place it on your whole.

☐ For example, you roll 1/4.

☐ The first player to cover their whole **exactly** wins.

Fraction Exchange Subtraction

☐ Start with your WHOLE covered with two halves.

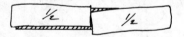

☐ Take turns rolling the die.

☐ Whatever you roll, you take off (or subtract) that fraction. You may have to exchange first. For example, if you roll 1/8 on your first roll, you must exchange 1/2 for 4/8 before you can subtract 1/8.

☐ The winner is the first player to uncover his or her WHOLE, exactly.

More Ideas

☐ Put two fraction kits together and play to cover up different amounts. For example, play to cover up two WHOLES, or one and one-half WHOLES.

☐ Play to see who can make the largest number after five turns.

Judy's Fractions

Why

To reinforce the understanding of fractions and mixed numbers

▶ **Mixed numbers** *are those that have a whole number and a fraction together.* ◀

How

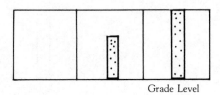

□ Each player takes the equivalent of 6 wholes out of his or her fraction kit, using wholes, 1/2's, 1/3's, 1/4's, 1/6's, 1/8's, 1/12's, and 1/16's.

□ At the beginning of each round, the players bet on whether the lead player will toss heads or tails.

□ The players decide together how much to bet for the round, say 2 1/4. Each player puts that amount of fraction pieces into the pot and announces his or her bet: heads or tails.

□ The lead player tosses the coin.

□ The lead player divides the pot evenly among the winners. The winners are responsible for checking that they were given the correct amount.

□ If an error was made, the lead player forfeits 1/4 extra for the next pot.

□ If the pot cannot be divided evenly among the winners, the extra pieces can be left to sweeten the next pot—or traded for smaller pieces (1/12's or 1/16's) that can be divided evenly. For example, if 1/4 is left over with three winners, trade 1/4 for 3/12.

□ Lead player passes to the left after each round.

□ Play continues for a specified number of rounds, say 5 or 10; or a specified time, say 5 to 20 minutes; or until one player has won all of the other players' fraction pieces.

Grade Level

TOOLS

Fraction Kit for each player (see pages 120-123)
Pennies

A game for 3–8 players

Place the Digits

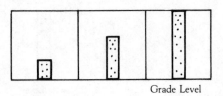

Grade Level

TOOLS

Pencils or pens

Paper

Dice, spinners, or playing cards

A game for 2–6 players

Why

To develop understanding of place value and number position relationships

▸ *This activity is a good introduction to the following missing digit activities, familiarizing children with opportunities to change the value of numbers by assigning digits to different positions.* ◂

How

☐ Use a die or spinner with the digits 0 through 9. (Or draw playing cards Ace through 10, counting Ace as 1 and 10 as 0, replacing cards as you go.)

☐ Players each draw three boxes on a piece of paper.

☐ ☐ ☐

☐ The leader spins or rolls a die and announces the number.

☐ Each person places that number in one of the three boxes.

☐ No changes are allowed after the digit is recorded.

☐ The leader spins or rolls two more times, and each time the players place the digit in any empty box.

☐ Each player reads his or her three-digit number. The player with the largest number wins.

☐ Repeat the game as many times as you wish, taking turns being leader.

More Ideas

☐ Play to get the smallest number, or to get as close as possible to a particular number, say 500.

☐ Allow one extra spin or roll and a reject box.

☐ ☐ ☐   ☐ Reject box

☐ Play with more boxes to make a four- or five-digit number.

☐ ☐ ☐ ☐ ☐

☐ Make fractional numbers.

☐ Use boxes that form an arithmetic problem. For example:

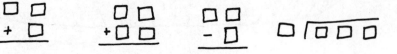

Find the Digits

Why

To practice identifying the value of numerals based on their position in a mathematical statement

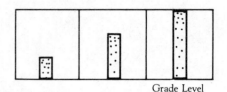

Grade Level

> ▶ *These activities contribute to understanding of place value and positional value.*
>
> $$\begin{array}{r} 85 \\ -37 \\ \hline \end{array}$$
>
> *For example, in the problem 85 − 37, or it is important for a child to recognize that the 7 must be taken away from the 5 and that it will be necessary to "borrow" so that the 5 becomes 15.*
>
> *Problems with missing digits are often used to check a child's understanding of the operations of addition, subtraction, multiplication and division, and may be found in some tests.* ◀

How

☐ Tell your children this story:

> One day Aunt Bebe greeted her young friends Sue and Arlene at the door with a long face. When they asked her what was the matter, she said that somebody had given her a nice sheet of arithmetic problems, all solved; and the most terrible thing had happened to it!! The dog had come in from a swim, given a big shake, and had splattered water all over the sheet. Many of the numbers had just melted away! She didn't know what to do, because the person who had given her the sheet had worked hard to get all those problems right. Of course Sue and Arlene offered to help her right away. Working together, they found all of the missing numbers in short order! Can you?

☐ Cut out the number squares, and have your family move the numbers around on the problems, so that if a number is in the wrong place it can be moved easily instead of having to be erased. Saves a lot of erasure holes in the paper!

TOOLS

Several sets of number squares, 0-9

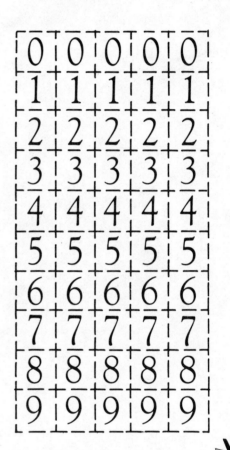

□ When all of the numerals are in place for a particular problem, discuss the reasons for putting them in those places. Here is an example:

$$
\begin{array}{r}
5\ 6\ \square \\
+\ 7\ \square\ 1 \\
\hline
\square\ 3\ 0\ 8
\end{array}
$$

□ What number goes in the top box?
(It has to be 7, since 7+1=8.)
□ What number goes in the box between the 7 and 1?
(It has to be 4, because 6+4=10.)
□ That means that the number in the bottom box will be 1, since 5+7+1=13. The **1** comes from the ten, carried into the next column.

□ Try these problems:

$$
\begin{array}{r}
1\ 4\ \square\ 0 \\
-\ 7\ 5\ 4 \\
\hline
7\ 1\ \square
\end{array}
\qquad
\begin{array}{r}
3\ \square \\
\times\ 5 \\
\hline
1\ \square\ 5
\end{array}
\qquad
\begin{array}{r}
2\ 8 \\
+\ 3\ \square \\
\hline
\square\ 1
\end{array}
$$

□ Have your children work together, with you or with each other, to solve the problems on the next page.

More Ideas

□ Have children make up missing digit problems. Write down any arithmetic problem, and draw boxes around some of the numbers (not too many at first!). Then rewrite the problem on another sheet of paper, with the boxed numbers left out.

Example:

$$
\begin{array}{r}
3\ 8\ 2 \\
-\ 7\ 4 \\
\hline
3\ 0\ 8
\end{array}
\qquad
\begin{array}{r}
3\ \boxed{8}\ \boxed{2} \\
-\ 7\ 4 \\
\hline
\boxed{3}\ 0\ 8
\end{array}
\qquad
\begin{array}{r}
3\ \square\ \square \\
-\ 7\ 4 \\
\hline
\square\ 0\ 8
\end{array}
$$

Have the children check that their problems are possible to do, then give them to each other to try. Do some problems have more than one solution?

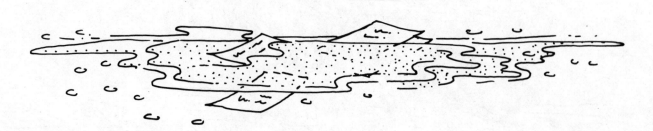

FIND THE DIGITS Addition and subtraction

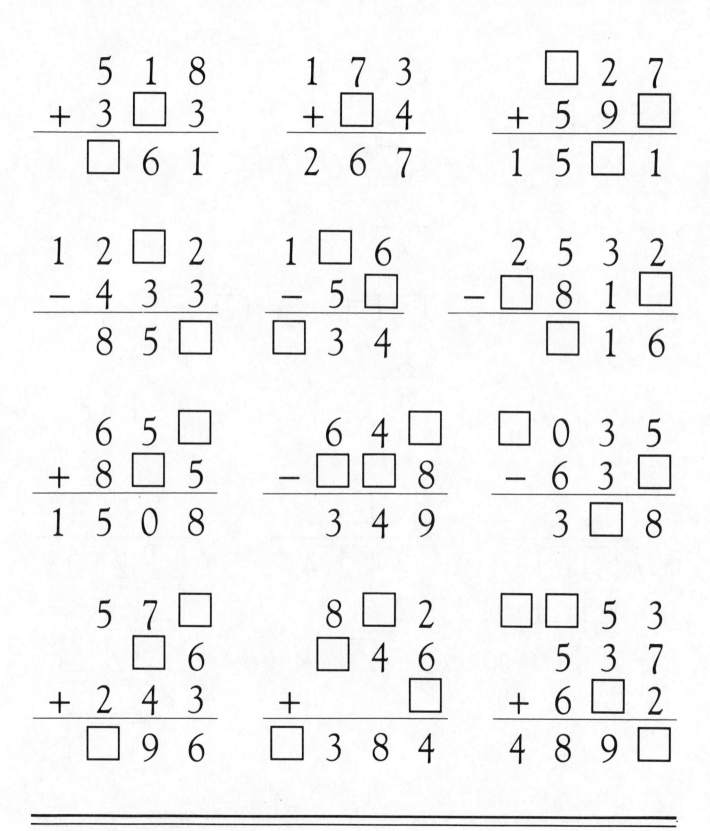

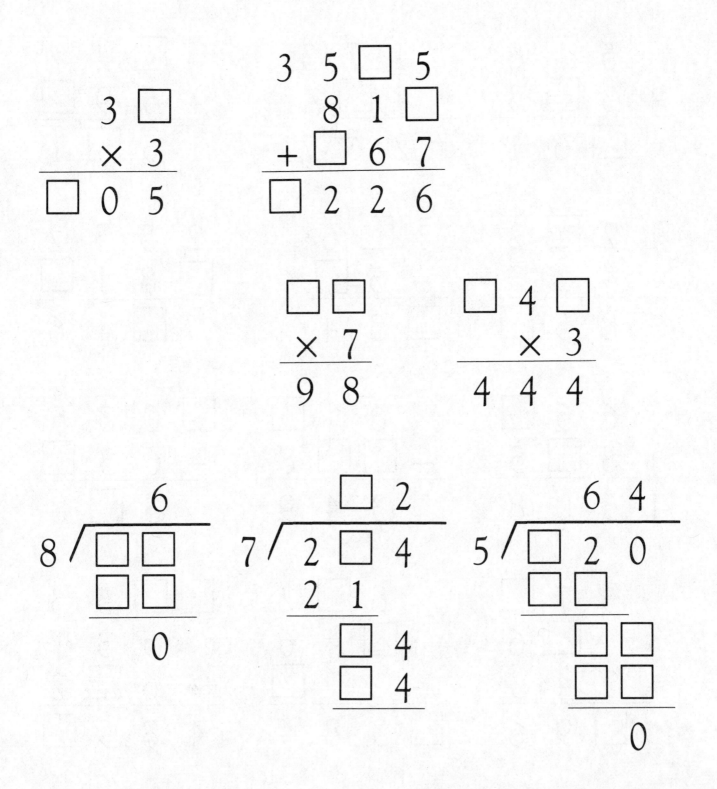

FIND THE DIGITS Addition and subtraction

Answers

```
    5 1 8          1 7 3          [9] 2 7
  + 3[4]3        + 9 [9]4        + 5 9 [4]
  ---------      ---------      -----------
   [8]6 1         2 6 7         1 5 [2] 1
```

```
  1 2[9]2        1[8]6          2 5 3 2
  - 4 3 3        - 5[2]        -[1]8 1[6]
  ---------      ---------      -----------
   8 5[9]        [1]3 4         [7]1 6
```

```
    6 5[3]          6 4[7]        [1]0 3 5
  + 8[5]5        -[2][9]8        - 6 3[7]
  ---------      ---------      -----------
  1 5 0 8         3 4 9         3[9]8
```

```
    5 7[7]          8[3]2        [3]7 5 3
    [7]6          [5]4 6          5 3 7
  + 2 4 3        +    [6]        + 6[0]2
  ---------      ---------      -----------
  [8]9 6         [1]3 8 4       4 8 9[2]
```

Answers

```
       3 [5]
      ×   3
     ─────────
   [1]  0   5
```

```
   3   5  [4]  5
          8   1  [4]
   +  [8]  6   7
   ────────────────
   [5]  2   2   6
```

```
      [1] [4]
     ×      7
     ─────────
       9    8
```

```
   [1]  4  [8]
         ×    3
     ───────────
       4   4    4
```

```
          6
      ───────
   8 / [4] [8]
      [4] [8]
      ────────
          0
```

```
         [3]  2
      ──────────
   7 /  2  [2]  4
       2   1
       ─────────
         [1]  4
         [1]  4
```

```
         6    4
      ──────────
   5 / [3]  2   0
      [3] [0]
      ──────────
         [2] [0]
         [2] [0]
         ──────
              0
```

☐ Have the children take three of the number squares, say the 1, 2, and 3.

Put the number squares together in every possible way, finding all the different arithmetic problems, and their answers, that they can make with those three numbers. Use addition, subtraction, multiplication, or division, as appropriate.

For example, these problems:

$$
\begin{array}{ccc}
1\ 2 & 1\ 3 & 3\ 1 \\
\times\ 3 & -\ 2 & +\ 2 \\
\hline
3\ 6 & 1\ 1 & 3\ 3
\end{array}
$$

will become these problems when rewritten:

Use 1, 2, and 3:

$$
\begin{array}{ccc}
\square\ \square & \square\ \square & \square\ \square \\
\times\ \square & -\ \square & +\ \square \\
\hline
3\ 6 & 1\ 1 & 3\ 3
\end{array}
$$

What other combinations can be made from 1, 2, and 3?

Use other numbers, such as 3, 5, and 7, to make up problems.

☐ Try making up some problems that use fractions, decimals, and percents.

For example
  10% of □ 50 = 15, or
  1/□ of 2□ = 8, or
  23.□ + 5.2 = □8.7

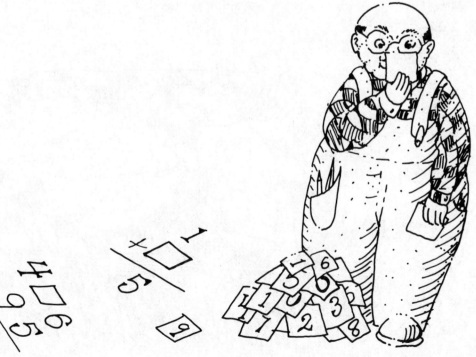

Three Bean Salads

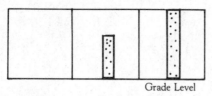

Grade Level

TOOLS

3 types of dry "beans"
Red beans
Lima beans
Black-eyed peas
Paper plates or paper cups to hold small portions of beans

Why

To practice working with **ratios and proportions**

▸ *The language of ratio and proportion is very important in present-day mathematics. A* **ratio** *is the numerical relation between two quantities, usually determined by dividing one of the numbers by the other and expressing the result as a fraction or a percent. For example, a business might determine its ratio of assets to liabilities by dividing the value of the assets by the value of the liabilities.*

▸ *In geometry, the ratio of the circumference of a circle to its diameter is always the number π or pi or about 3.1416.*

▸ **Proportion** *is a statement showing that two ratios are equal. For example, the ratio 1/2 is the same as the ratio 3/6 or 2/4. This is an important idea in algebra, since if any three of the numbers in a proportion are known, the fourth can be found—this is the "unknown" in algebra problems.* ◂

The following section includes fairly difficult algebra problems, which can be solved easily by trial and error using the beans.

How

□ **All three types of beans go into each salad.**

□ Children should be encouraged to guess and adjust as they work. Use the beans to solve the problems.

□ For each salad determine how many of each of the three types of beans are needed.

THREE BEAN SALADS

Each salad contains Red beans, Lima beans, and Black-eyed peas

1

This salad contains:

2 Lima beans

Twice as many Red beans as
Lima beans

10 beans in all

2

This salad contains:

4 Red beans

1/2 as many Black-eyed peas as
Red beans

10 beans in all

3

Lima beans make up 1/2 of this salad:

The salad has exactly 2 Red beans

The number of Lima beans is double
the number of Red beans

4

This salad contains:

The same number of Red beans
as Lima beans

3 more Black-eyes than Red beans

A total of 18 beans

5

This salad contains 12 beans

1/2 of the beans are Red

Lima beans make up 1/4 of the salad

6

This salad contains at least 12 beans

It has one more Lima bean than
Red beans

It has one more Red bean than
Black-eyes

7

This salad contains:

3 times as many Red beans as
Black-eyes

One more Lima bean than Red beans

8 beans in all

8

This salad contains:

An equal number of Red beans and
Black-eyes

5 more Lima beans than Red beans

No more than 20 beans

Make up a different salad.

Write instructions for someone else to make your salad.

Gorp

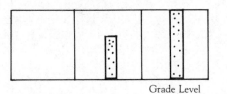

Grade Level

TOOLS

Gorp worksheet
3 dice
Pencil

A game for
2 or more players
or teams

Why

To develop understanding of ratios and proportions

▶ *This activity includes experiences with probability as well as planning ahead—and of course, some luck for success!* ◀

How

☐ Play the first game as a group.

☐ Read the Gorp worksheet directions to your family:

> A camping store makes trail mix from peanuts, raisins, banana chips, dried apricots, and walnuts. Each bag can have three ingredients. You will throw three dice to find out how many ounces of each of the three ingredients is in every bag. You will decide how many bags of these ingredients to make. You want to use up as much as possible of your ingredients in your rounds of play. You are penalized for the cost of the leftovers.

Further instructions:

☐ Each of you will start with $50 worth of Gorp materials: 100 ounces of peanuts, 100 ounces of raisins, 100 ounces of banana chips, 50 ounces of dried apricots, and 50 ounces of walnuts.

☐ On each turn, you will roll three dice to determine quantity of ingredients for your bags of Gorp.

☐ Each bag will have only **three** of the five possible ingredients.

☐ Your goal is to finish the game, after four turns, with little or no leftover ingredients. The person who has leftovers with the smallest value wins by having the smallest "loss" from unused materials.

☐ The dice throw determines how many ounces of each ingredient you will pack into your bags for that round.

☐ You will choose: which three ingredients to use, and how many **whole** bags to make on each round.

☐ No fractional parts of bags are allowed.

☐ You may not use more of any ingredient than you started with.

☐ If you cannot package at least one whole bag on a round, you must record **zero** bags for that round.

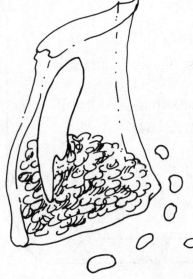

□ Here is an example of a first round:

　□ Suppose the dice show 2, 3, and 5. Record these numbers next to "1. dice throw: ____ ____ ____." You may use two ounces of one ingredient, three of another, and five of a third ingredient to pack into the bags on this round.

　□ You may choose any three of the ingredients, as long as you have enough of them.

　□ Suppose you decide on two ounces of walnuts, three of apricots, and five of raisins.

　□ To keep track, record the appropriate initials over the 2, 3, and 5.

　□ Now you may choose how many bags of this mixture you will make. Suppose you decide on eight.

　□ Calculate how much of each ingredient you use by multiplying 8 times the number of ounces you are using per bag.

　□ The results are recorded on the Sample Work Sheet. Subtract what you used, to get ready for the next round.

	Peanuts	Raisins	Bananas	Apricots	Walnuts
You start with this many ounces:	100	100	100	50	50
Round 1 Dice throw *W A R* 2 3 5					
_8_ bags	—	40	—	24	16
Left over	100	60	100	26	34
Round 2 Dice throw __ __ __					
____ bags	___	___	___	___	___
Left over	___	___	___	___	___
Round 3					

□ On your next turn, you will decide again the amounts of any ingredients you will use (as long as there are enough) and proceed in the same manner.

□ After four turns, your goal is to have no leftovers or have to forfeit the smallest possible price for your leftovers.

□ If your children do not seem quite ready to play on their own, continue to play as a group.

GORP GAME WORKSHEET

★ A camping store makes trail mix from peanuts, raisins, banana chips, dried apricots, and walnuts. Each bag can have three ingredients.

★ You will throw three dice to find out how many ounces of each ingredient is in every bag.

★ You will decide how many bags of these ingredients to make.

★ You want to use up as much as possible of all your ingredients in four rounds of play.

★ You are penalized for the cost of the leftovers.

	Peanuts	Raisins	Bananas	Apricots	Walnuts
You start with this many ounces:	100	100	100	50	50

Round 1

Dice throw __ __ __

_____ bags   _____ _____ _____ _____ _____

Left over   _____ _____ _____ _____ _____

Round 2

Dice throw __ __ __

_____ bags   _____ _____ _____ _____ _____

Left over   _____ _____ _____ _____ _____

Round 3

Dice throw __ __ __

_____ bags   _____ _____ _____ _____ _____

Left over   _____ _____ _____ _____ _____

Round 4

Dice throw __ __ __

_____ bags   _____ _____ _____ _____ _____

Left over   _____ _____ _____ _____ _____

For your score, find the price of what is left.

Peanuts _____ oz. at 10¢ _____

Raisins _____ oz. at 10¢ _____

Banana chips _____ oz. at 10¢ _____

Apricots _____ oz. at 20¢ _____

Walnuts _____ oz. at 20¢ _____

Total cost of leftover ingredients _____

Chapter VI

PROBABILITY
and
STATISTICS

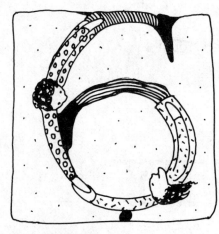

TOOL KIT

Large paper for making graphs

Smaller sheets of colored paper

Scratch paper

Graph paper (see page 79-82)

Pens and pencils

Scissors

Glue and tape

Dice

Cube-shaped blocks

Cardboard for making spinners

Ruler

Paperclips

Beans

Checkerboard

Game markers

3″×5″ cards

PROBABILITY AND STATISTICS

Probability and statistics are two of the most useful branches of mathematics for daily life. They help us understand the world around us and organize information in ways that make it easier to comprehend.

For example, if the Shapely Shirt Company wanted to know how long to make the sleeves on their shirts, but didn't know how long people's arms were, they would apply probability and statistics. It would be an impossible task to measure **every**one's arms, but a sample of people could be measured, and **probably** most of the rest of the shirt buyers would have arms about the same length with about the same distribution.

First, the company would send someone out to talk to many people. The surveyor would probably work with a tally sheet something like this:

26	27	28	29	30	31	32	33
	/	//			/	/	++++

After the survey, a statistical graph would be prepared to present the information. It might look like this:

ARM LENGTH (in inches)	X =10 people
26	XX
27	X
28	XXXXXXX
29	X
30	
31	X
32	XX
33	XXXXXXXXX
34	
35	X

In reality, a company would probably survey many more people than shown here, but for explanation purposes we have kept the numbers small.

This is only one of many examples. We might want to know the probability of winning the school raffle or the chance of rain tomorrow. Space scientists have to calculate the probability that a space shuttle will not collide with a meteor or other object in space. Businesses are constantly reviewing statistics to look for patterns and predictions.

Probability and statistics traditionally have not been a very large part of elementary and middle school curriculum. Many educators feel, however, that these topics are among the most important for children to understand and that they are an easy and effective way to learn and use arithmetic with high interest for students.

In this section, we have included activities that develop understanding of the basic concepts of probability and statistics, without dwelling on complicated terminology. We do introduce a few important vocabulary words, such as **mean, median,** and **mode,** where they are appropriate and where the explanation of their meaning can be included in the context of their use. To give you a little head start on the definitions:

☐ **mean** is the same as **average.** For the shirt sleeves above, it would be obtained by first multiplying the number of people in each category by that sleeve length, adding all of those answers, then dividing the total by 250, or the total number of people surveyed. The mean here is 30 inches.

☐ **median** is the number that would fall in the middle if the results were arranged in order from smallest to largest. In our arm-length graph, the median is the length of 31 inches. An easy way to find this is to cross off one X at the bottom, then one X at the top. Keep crossing off top and bottom until there is only one mark left. That's the middle or median number.

☐ **mode** is the item that occurs most frequently. We can see that for the shirt sleeve lengths, there are two modes, since both the 28-inch length and the 33-inch length have more X's. This is called a **bi-modal distribution.**

Each of these ideas is important for different purposes. The **mean** would be useful in calculating the number of yards of shirt materials to be purchased. The **median** might be important in deciding how to set the fabric cutting machines. The **mode** would certainly be essential in deciding which length people would be most likely to buy.

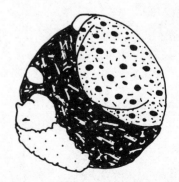

GRAPHING

The graph we have used is called a **bar graph,** since the results look like bars. There are many other kinds of graphs, such as circle or pie graphs, picture graphs, line graphs, and so on. Some of these are illustrated in the following activities.

Graphing is a way to organize data, to record information, and to provide easily understood answers to questions. Graphs may simply compare two groups, or they may depict complex situations. The steps in graphing:

☐ Decide on a question to ask

☐ Think about what the outcome might be.

☐ Gather the data.

☐ Organize the information to put onto a graph.

☐ Display the information on the graph.

☐ Interpret the graph, with statements or questions.

Graphs can differ in the way they show information:

   Real graph—uses real objects, such as toys, fruit, shoes.

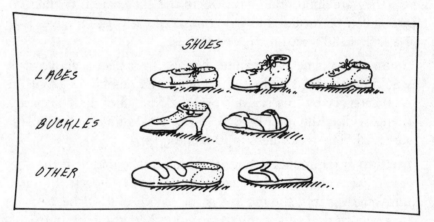

Picture graph—uses pictures or models to stand for the real thing.

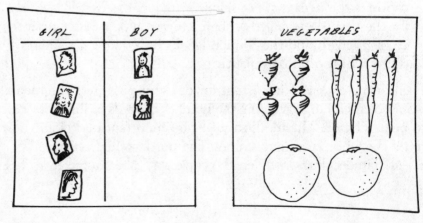

Symbolic graph—uses symbols to stand for real things.

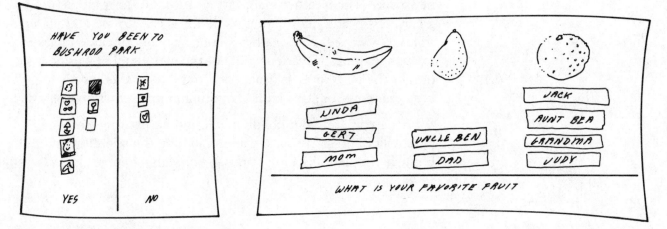

Graphs can also differ in their format:

Circle graphs

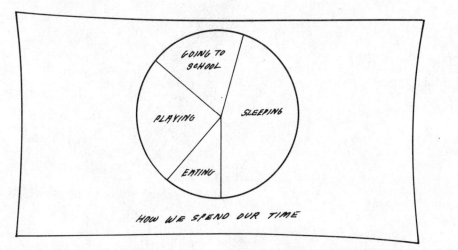

Line graphs

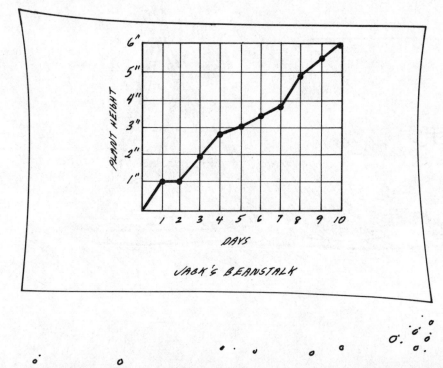

An excellent place to watch for tables and graphs is in the daily newspaper. The weather map, for instance, is a form of picture graph, organizing information so that it is easy to see and understand at a glance. The weather section may also include tables of the temperatures in various parts of the country. The business section of the newspaper is usually full of tables and graphs, since business people use this kind of information presentation every day.

We hope you and your children will find an exciting world of probability and statistics to explore, joining the scientists and business people in looking at things through "graphic glasses."

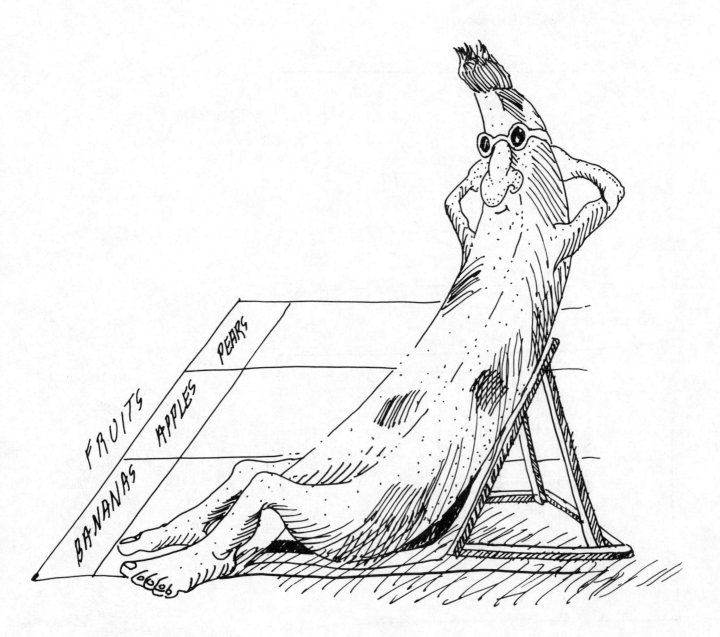

Graphing Information

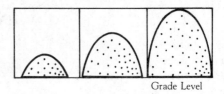
Grade Level

Why

To practice gathering information and organizing it into graphic format

How

- ☐ Help your child choose one of the topics on the next page (or make up one of your own) to use in a survey.
- ☐ Discuss how you both will collect information: Who will you each ask? How will you keep track of the answers? How will you make sure the same person isn't asked twice? and so on...
- ☐ Then collect answers to your question. You should probably try to survey at least ten people or more if that is possible.
- ☐ Make a graph to represent the information collected. Use one of the kinds of graphs shown on pages 142 and 143, or invent a new kind of graph. Use large pieces of paper to display the graph. Use felt pens or paper cutout shapes to show the tally. Your graph should be easy to understand and interesting to look at.
- ☐ Ask each person who looks at the graph to write a sentence about it. Turn the sentences into questions, and put them on the graph. For example, if you had made a graph about favorite fruit, somebody might write a sentence, "There were more bananas chosen than apples." A matching question might be "How many more bananas were chosen than apples?"
- ☐ Ask other people to see if they can answer the questions on your graph.

▶ *A math teacher in Santa Barbara, California, named Hal Saunders,* made a survey of one hundred people in different jobs. Over two-thirds of the people said they used statistical graphs at work. Children need to be able to make, read, and interpret graphs to become informed consumers and citizens.* ◀

More Ideas

- ☐ Graph your information in several ways.
- ☐ Find the mode, mean, and median (if appropriate) for your graph. See the chapter introduction for definitions.
- ☐ Look in newspapers and books to find different kinds of graphs. Make a scrapbook of all the graphs you can find, so that when you are looking for a graphing idea, you will have some samples.

*see bibliography

TOOLS

Large paper
Pens
Pencils
Colored paper
Scissors
Glue

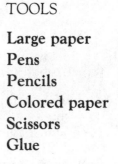

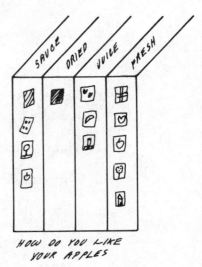

HOW DO YOU LIKE YOUR APPLES

SOME SURVEY TOPICS

1. How many brothers do you have? How many sisters?

2. How far do you travel to go to school?

3. What is the first letter of the name of the street you live on?

4. Do you live on a street, avenue, boulevard, lane, road...?

5. In what state or country were you born?

6. How many letters are there in your first name? Your last name?

7. What is your favorite color?

8. What is the length of your left thumbnail?

9. What is your shoe size?

10. What day of the week were you born? (Look in the telephone directory for a calendar, or call the public library.)

11. What is the next to last letter in your first name? Your last name?

12. What would you like to be when you grow up?

13. What is the last digit of your telephone number?

14. What is the third digit of your telephone number?

15. What is your favorite ice cream flavor? (You may want to ask people to choose between vanilla, chocolate, and strawberry.)

16. Who is your favorite singer?

17. What kind of pet do you like best?

18. What time did you go to bed last night?

19. What did you have for breakfast?

20. How many aunts and uncles do you have?

The Means Justify the End

Why

To develop understanding of the kind of average called a **mean**

> ► *A* **mean** *is found by adding all the measurements or numerical values you have and dividing by the number of values. Most people are referring to a mean when they talk about "the average," although* **average** *may be correctly used to indicate the mean, the median, or the mode.* ◄

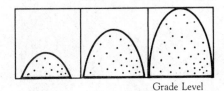

Grade Level

TOOLS

One die

Cube-shaped blocks or paper squares

A sheet of paper for each person

How

- ☐ The object of this activity is to find the **mean** of five rolls of a die for each person.
- ☐ Each person rolls the die on his or her turn, and takes the number of blocks or paper squares indicated on the die.
- ☐ Put the blocks or squares in a row, beginning at the edge of the paper. Each roll of the dice should be in a separate row, as shown.
- ☐ Continue taking turns rolling the die until everybody has had five turns and has made five rows.
- ☐ To find out each person's **mean,** even out the lengths of the rows by moving the blocks or squares from one row to another. Be sure to keep five rows.
- ☐ If there are leftovers, keep them slightly apart.
- ☐ What is the **mean** number rolled by each person? Are the numbers very different or close together?
- ☐ Move all the rows together (for example, if there are three people, putting the rows together will make fifteen rows. Move the blocks again to make all of the rows even. Is the answer very different?

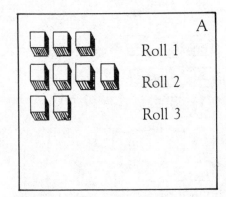

Roll 1

Roll 2

Roll 3

A

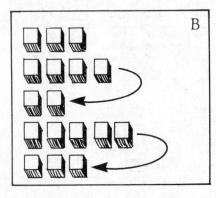

B

More Ideas

- ☐ After older students have tried this activity with blocks to get a visual picture of a mean, have them play without blocks, using a pencil and paper to record their five rolls. To determine the mean, add the five rolls and divide by five. The person with the largest quotient wins.
- ☐ Or vary the rules so that the person with the smallest remainder wins.

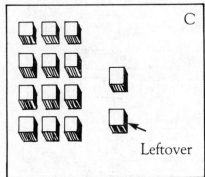

Leftover

C

How Long Is a Name?

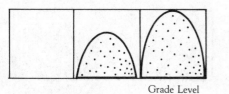

Grade Level

TOOLS

1″ squares
Pencil
Paper
Glue

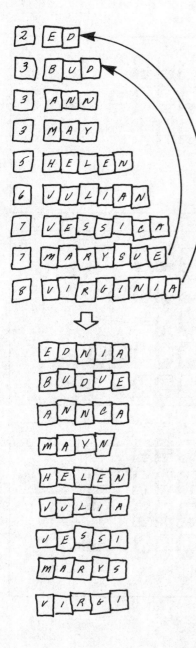

Why

To introduce the statistical concepts of mean, median, and mode, and to provide practice in making a bar graph

How

☐ Make a list of the names of your family and some relatives or friends.
☐ Write the letters of each name on the 1″ squares, using one square for each letter.

☐ Write the number of letters in each name, and the person's initials, on another square.

☐ Line up the names from longest to shortest, as shown in the picture.

Mean

☐ Find the average that is called a **mean** of the lengths of the names. To do this, move letters from the longer names to fill in the shorter ones, until all the rows have the same number of letters. (It doesn't matter where the letters go, as long as the rows have the same number of letters, or as close as possible.)
☐ The **mean** in our example is a little less than five, because all the names evened out to be five letters long, except one.

Median

☐ Now put out the squares with the numbers that tell how long each person's name is. Arrange them in numerical order:

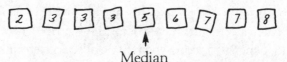

Median

☐ Find the center number in the row. This is the **median.** In our example, "5" is in the middle, so **five** is the median for this example. If there are two numbers in the middle, add them together and divide by two to compute the median.

Mode

☐ Next, glue all of the numbers onto a bar graph like the one shown here. Look for the number which occurs most often. This is called the **mode**.

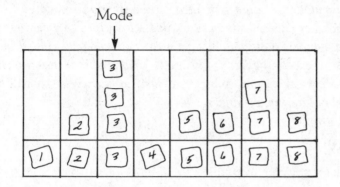

Mode

To summarize, our sample group has:

a **mean** name length of 4.8,

a **median** of 5, and

a **mode** of 3.

☐ What happened with your group of names?

More Ideas

☐ Talk about the difference between means, medians, and modes. Why don't they all come out the same?

☐ Can all graphs have means and medians?

☐ Can there be a graph that has the same number for the mean, median and mode?

Fair Spinners I

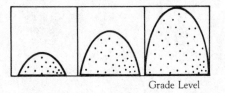

Grade Level

TOOLS

**Spinners with colors
(see directions on
page 154)**

Pencil

Paper

Why

To do physical experiments that develop an intuitive understanding of probability theory

▶ *Accumulation of data generated by the spinners simulates the kind of probability experiments done by real scientists. The experiments produce numbers that gradually get near the theoretical values, but rarely attain the values exactly. This quirk of probability theory often confuses high school and college students. Early experiences will help develop later understanding.* ◀

How

☐ Make a spinner like this: (See directions on page 154.)

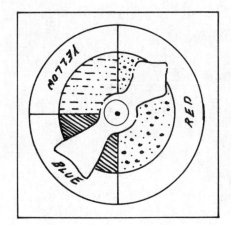

☐ Spin your spinner 36 times.
☐ Keep track of which colors occur.
☐ Make a graph to show the results.

RED	BLUE	YELLOW
		X
		X
X	X	X
X	X	X

□ Now, study the spinner. What portion of it is red?____
blue?____ yellow?____

How does this compare with your graph?

► *Mathematicians say that with a **fair** spinner, one that spins evenly, with divisions like yours,*

    the probability of spinning red is 1/2

    the probability of spinning blue is 1/4

    the probability of spinning yellow is 1/4.

*This means that, in 36 spins, you would expect to get half of 36 or 18 reds, one-fourth of 36 or 9 blues, and one-fourth of 36 or 9 yellows. These numbers are what you **expect** in **theory**. However, an experiment will rarely produce numbers that match the expected numbers exactly.* ◄

□ Now spin and record your spinner 36 more times, or combine your results with those of somebody else who has done the same spinner. How does this graph compare with your original graph?

► *To find the expected numbers, red is half of 72, blue is one-fourth of 72, and yellow is one-fourth of 72.*

► *Are you closer with the combined data from more spins than you were with 36 spins?*

► *If you were to keep on spinning for a very long time, your results would get closer and closer to the expected numbers, provided your spinner is fair. This reflects what is called **The Law of Large Numbers** in probability theory.* ◄

More Ideas

□ For younger children, omit the discussion of probabilities and the calculation of the expected numbers. Instead, talk about whether their results are about half red, one-fourth blue, and so on.

□ Make spinners that have different arrangements of colors, such as 1/3 red, 1/2 blue, and 1/6 yellow.

□ Just for fun, make a whole bunch of beautiful spinners and see how many you can get to spin at once.

Fair Spinners II

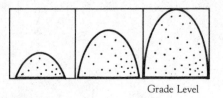

Grade Level

TOOLS

Spinners with numbers (see page 154)

Graph paper

Pencil

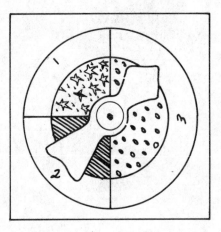

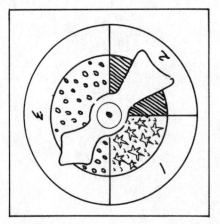

Why

To explore probability by performing experiments with spinners

How

Matching Numbers

☐ Spin two spinners or one spinner twice. (See page 154 for directions for spinners.)

☐ Record the numbers you get. Do this 25 times.

☐ Make a graph of how many times you get:

a 1 on both spinners,
a 2 on both spinners,
a 3 on both spinners.

Both 1's	X X
Both 2's	X
Both 3's	X X

☐ Look at the area on the spinner for each number. Which number would you expect to have more matches? How does this compare with your results? Repeat the experiment (or combine your results with those of someone else) and look at the results. How do the combined results compare to what you would expect?

Spinner Sums

☐ Use two spinners. Look at the spinners and predict the sums you will get if you spin both spinners and add the results together. If you were to spin 25 times, what sum or sums do you think would occur most often?

☐ Spin 25 times and make a graph to show your results.

☐ Are the results different from your prediction?

► *In probability, we make lists of all possible results of an experiment to help us find the chance of a certain event happening. Because your spinner has twice as much area for the 3, we might think of its numbers as 1, 2, 3, and another 3 or 3a.*

▶ *We can then make the following list of all the possible combinations.*

First Spinner	Second Spinner	Sum
1	1	2
1	2	3
1	3	.
1	3a	.
2	1	.
2	2	
2	3	
2	3a	
3	1	
3	2	
3	3	
3	3a	
3a	1	
3a	2	
3a	3	
3a	3a	

▶ *So we see we have sixteen possible pairs of numbers that can occur.*

▶ *How many of these are matched 1's? _____*
2's? _____ 3's? _____

▶ *Now fill in the sums in the list of possible spins. There is only one sum of 2, so we say the probability of spinning a sum of 2 is one out of sixteen or 1/16. Find the probability of spinning sums of 3, 4, 5, and 6.*

▶ *How do these theoretical probabilities compare with your results in the **sums** experiment?*

▶ *Make a spinner with more numbers to use for these same experiments.* ◀

Making Spinners

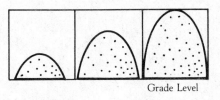

Grade Level

TOOLS

Cardboard
Scissors
Ruler
Pencil
Paper clip
Tape

How

☐ Cut out a cardboard arrow shaped like this:

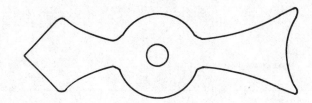

☐ Punch, with punch, hole in center.
☐ Cut a scrap for a paper washer into a square and punch a hole in center with punch.

☐ Cut out a four-inch square of cardboard.
☐ On the four-inch card, measure two inches along on all sides and mark lightly:

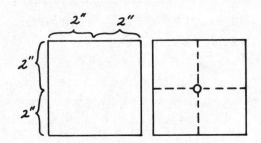

☐ Connect these marks with faint pencil lines.
☐ Mark center with a dot.
☐ Make hole in the center of the spinner card with thumbtack or end of paper clip.
☐ Draw a design for the activity you want to do:

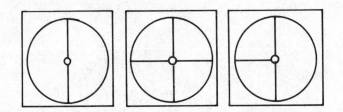

☐ Put bent paper clip through center hole.

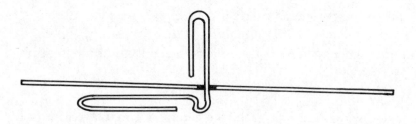

☐ Put your square and then spinner onto the spinner card and then clip down.

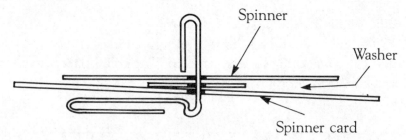

☐ Put masking tape on the bottom to hold paper clip.

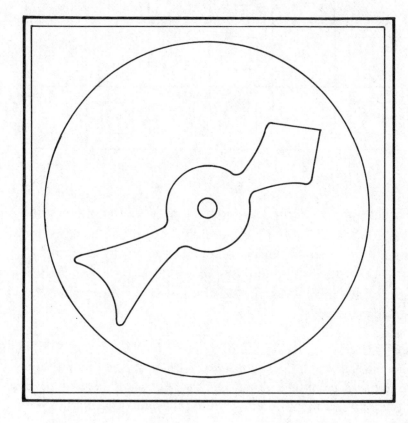

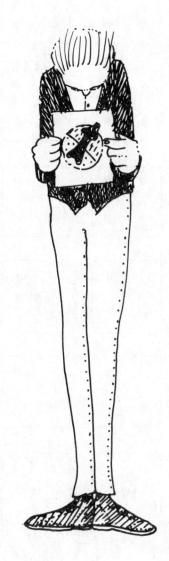

Rolling Records—Step I

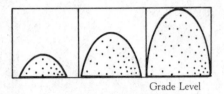
Grade Level

TOOLS

Dice (or spinners)
Graph paper
  (see page 79)
Dice squares
Glue

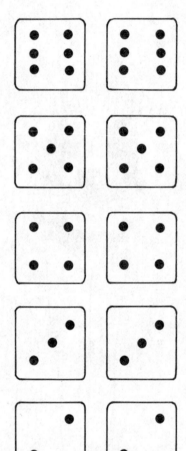

Why

To perform experiments that will help develop a sense of the probability of events that are about equally likely to happen, and to practice making data displays in the form of a bar graph

How

☐ Take one die. Look at each face carefully. Draw a picture of the dots on each face. If you were to roll the die 25 times, which number do you think would turn up most often? Write down your guess.

☐ Roll the die 25 times. Record each roll.

☐ Make a graph of the results. Make as many dice squares as you need (the same size as the squares of your graph paper) to show each number you rolled. Cut out the squares and glue them onto the graph.

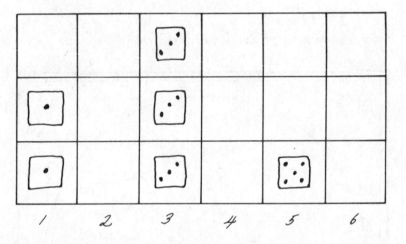

☐ Which number turned up most often? Were the others close?

☐ Roll the die 25 more times, and compare the results. Ask other family members to do the same. Are the results different?

☐ Put all of the results onto one graph. Has the shape of the graph changed? What do you think would happen after a great many rolls?

▶ *NOTE: ROLLING RECORDS, STEP 1, STEP 2, and STEP 3 are intended to be done as a series, in the order shown. The three activities develop understanding of the relationship between probability theory and the results of physical probability experiments.* ◀

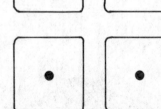
Dice squares

Rolling Records—Step II

Why

To develop, through dice experiments, an understanding of the probability of events that are not equally likely to occur

How

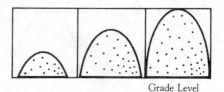

Grade Level

TOOLS

Dice
Graph paper
Pencils

□ Look at two dice together. Roll the dice once. What is the total number of dots? Make a list of all the different totals or sums that can turn up if you roll two dice together.

□ Roll your dice 36 times. Make a graph of your results. Mark an X for each sum you roll.

					X				
			X		X	X			
2	3	4	5	6	7	8	9	10	11	12

□ Look at your graph. Are you surprised at the results?

□ How do the results compare with the graph you made when you rolled one die?

□ Roll the dice 36 more times, or ask other family members to roll them 36 times.

□ Put all of the results onto the graph.

□ Has the shape of the graph changed? If so, how?

Rolling Records—Step III

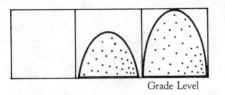

Grade Level

TOOLS

Dice

Graph paper
(see pages 79-82)

Why

To explore the theory of what **should** happen when you roll two dice and add

▶ *Even after many rolls of the dice, we rarely will get exactly what the theory predicts, but we should come closer and closer to those numbers if our dice are fair.* ◀

How

☐ Make a chart like that below.

☐ Pretend you are rolling one red and one blue die.

☐ The chart will show us all the possible combinations we can get with these dice.

☐ The numbers across the top are the possible numbers for the blue die.

☐ The numbers down the side are the possible numbers for the red die. For example, a blue 2 and a red 3 gives a sum of 5. This number has been recorded for you.

☐ Complete the chart.

▶ *When rolling two dice, some sums occur more frequently. Mathematicians say these are more likely to occur than the others.* ◀

BLUE / RED	1	2	3	4	5	6
1						
2						
3		5				
4						
5						

☐ Make a list of how many 2's, 3's, 4's, etc. are inside the chart.

▶ *Because there are 36 different sums on the chart, and six of them are 7's, mathematicians say the theoretical probability of rolling a seven is 6/36 or 1/6. This means that, in an experiment with fair dice, you would expect **about** six out of 36 sevens. How many 2's would you expect? 3's? other numbers?* ◀

☐ How do these numbers compare with your results in Rolling Records—Step II?

☐ Out of 72 rolls we would expect **about** two of the 2's, four of the 3's, six of the 4's, and so on. How do these numbers compare with your 72 rolls?

More Ideas

☐ How many rolls do you think it will take to **roll a double?**

☐ Roll until you get a double.

☐ Record how many rolls it took on a graph.

☐ Have other people roll for a double 50 times and record the results on the graph.

☐ Look at the graph. Are you surprised at the results?

☐ Study the probability chart. How often should you expect to get a double? Find the average number of rolls all of you took to get a double. (Add all of the results and divide by 50). Is this close to what you would expect? Discuss why the graph has the shape it does.

Raindrops

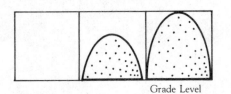

Grade Level

TOOLS

Raindrop board
  (see page 162)
2 dice of different colors
36 small beans

Why

To observe the relationship between theoretical probability and actual experimental results

▶ *This activity illustrates an interesting pattern of results from rolling two dice. The dice are rolled exactly 36 times, and 36 beans or "raindrops" are placed on the addition table for two dice, shown below. Each individual experiment will produce a very uneven pattern, but when you combine many repetitions of the experiment the pattern will tend to even out.* ◀

How

□ Look at the Raindrops Board. It is a table of the possible results of rolling two dice of different colors.

□ You will be taking turns rolling the two dice and putting a bean in the appropriate place on the table.

For example, if you roll a 2 on the blue die and a 3 on the red die, a bean will be placed on the 5 as shown here:

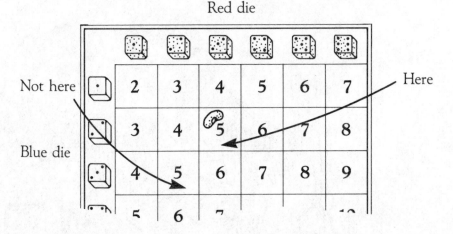

Be careful not to put the bean on the square that is the intersection of the red 2 and the blue 3—that is a different 5.

□ Before you begin, look at the table, think of what you know about probability, and ask yourself whether there will be a bean in every square. Will there be some boxes with no "raindrops," some with two or three?

□ Continue rolling the dice until all 36 beans have been used, then look at the pattern. Did it turn out the way you expected it to?

☐ Make a chart of how many boxes of no drops, one drop, two drops, three drops, and four or more drops.

ROUND	NO DROPS	1 DROP	2 DROPS	3 DROPS	4 OR MORE DROPS
1	10	15	6	5	0
2					
3					
4					
AVERAGE					

☐ Repeat the experiment several times, keeping a record of the results.

☐ After several games, take the average of each column.

► *If the dice are fair and you have repeated the experiment many, many times, the averages should be close to these: no drops—13, 1 drop—13, 2 drops—7, 3 drops—3, 4 or more—1.*

► *These numbers are related to the reciprocal of the number e (2.718), which is used as the base of the system of natural logarithms and which was made famous by the Swiss-German mathematician Euler. A reciprocal is the complement or counterpart of a number. For example, the reciprocal for 7 is 1/7, because $7 \times 1/7 = 1$ or $1 \div 1/7 = 7$.*

► *At first glance, this activity seems to lead to a paradox. One would expect the chance of landing in each box to be the same, but you probably found that over a third were missed in each game. If you were to continue rolling the dice many times and keeping track of the times you roll each of the 36 number pairs, you would find that the pattern would even out.*

► *One way to make sense of what is happening is to think of what happens when you roll a single die and keep track of the numbers you roll. You will not very often get 1 one, 1 two, 1 three, 1 four, 1 five, and 1 six, but if you rolled the same die many times, the results would move closer to being the same for each number.* ◄

RAINDROPS BOARD

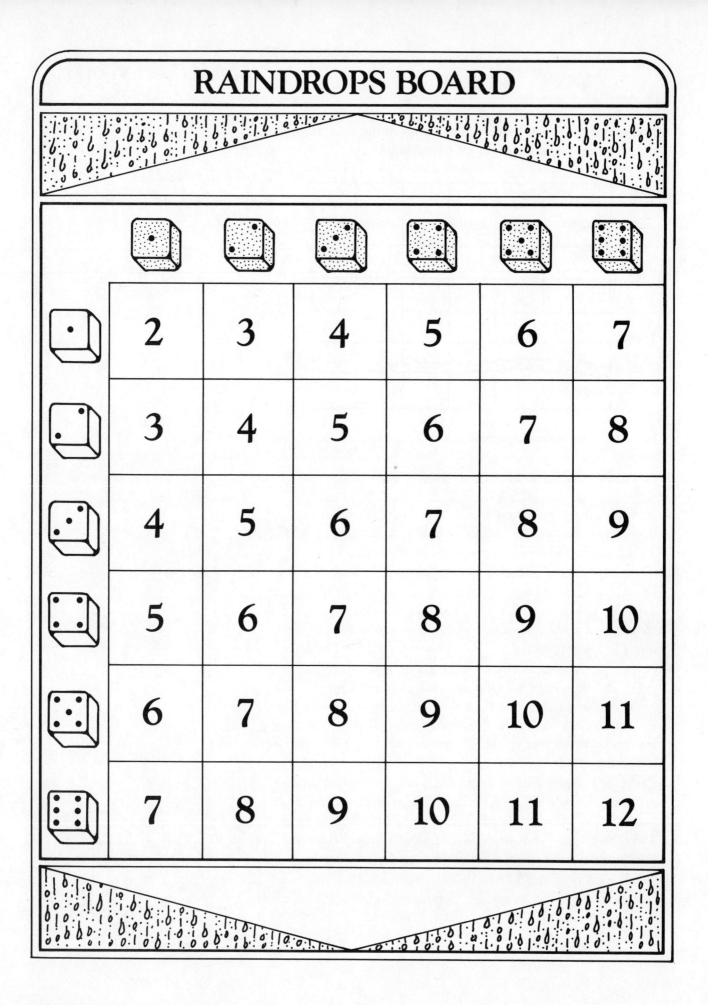

Random Walk I

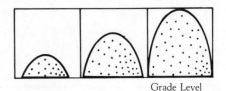

Grade Level

Why

To develop understanding of the ideas related to **random**ness, an important concept for both probability and statistics

▶ *Statisticians, when taking a poll to get information, take a **random** sample of the population, to avoid asking the question of only young people, or rich people, or some other special group. Randomness is also an important science concept, since air particles, for example, move at random, changing directions whenever they hit an obstacle.* ◀

TOOLS

Checkerboard
Spinners
  (see page 154)
Marker for each player
Masking tape

*A game for
2-4 players*

How

☐ Make a spinner that looks like this:

☐ Put small pieces of masking tape on a checkerboard as shown, and mark the points for North, South, East, West, and Center. The board represents a map of a fictional city. The streets of the city are represented by the lines of the checkerboard, and players will move along the lines.

☐ All players start in the center of the city, in the Start circle, facing North.

☐ Players take turns using the Random Walk spinner.

☐ Each player spins, turns his or her marker to face in the indicated direction, and moves one space along the line, in that direction, to the next intersection.

☐ Two or more players may occupy the same intersection.

☐ The goal is to be the first person out of the city—or off the board.

☐ Continue to play until all players are out of the city.

☐ After each person goes off the board, double the number of spaces for each move, so that if one person has gone off, the remaining players move two spaces each time. If two players have gone off, the remaining players move four spaces each time.

Random Walk II

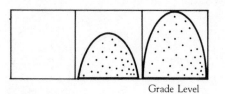

TOOLS

Random Walk board
(see page 166)
One die
Marker for each player
3″×5″ cards

*A game for
2-4 players*

Why

To practice generating random numbers and using simple compass directions to move around on a map

▶ *This game continues experience with randomness, adding possible strategies that will help control the results.* ◀

How

☐ Make a direction card on a 3″×5″ card or other small piece of paper. This card will determine the direction each player will move on his or her turn.

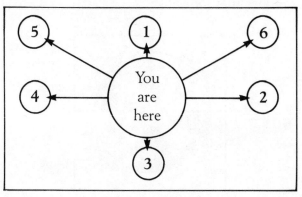

Direction card

☐ All players start at the center of the Random Walk board.

☐ The goal is to try to land on a high number at the end of the game, or to go off the board for 150 points.

☐ On each turn, the player **first** rotates the direction card to face so that the number 1 is toward the North, East, South **or** West. This **and** the number rolled on the die will determine the direction moved for that turn. In other words, imagine the marker is at the circle marked "You are here."

☐ Roll the die, and move in the direction indicated for that number.

☐ Take turns until everyone has had six moves.

☐ The score for each player is the number at the point where his or her marker is after the sixth roll.

☐ Any player who goes off the board in **exactly** six moves receives 150 points, but **none** if he or she goes off sooner.

☐ After the first game, discuss how you decide to place the direction card.

☐ Play five rounds of six rolls. The player with the highest total score wins.

More Ideas

☐ For younger children, mark out a large grid for them to actually walk on, without the numbers. Make a spinner that looks like this: Proceed with the same rules, except that the object is to be the first person out of the grid area.

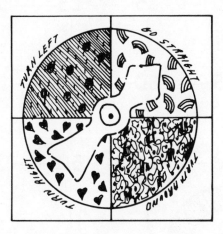

☐ For older children, make a spinner like the one shown and fasten **onto the game board.** Proceed with the same rules, with the object being to go off the board. Note that this version will be trickier, because it is hard to picture yourself in position on the board.

☐ Take a walk with your family as one team. Decide which way to go at each corner, using the "go straight/turn right" spinner. How far away from your start do you think you will be after five spins?

RANDOM WALK BOARD

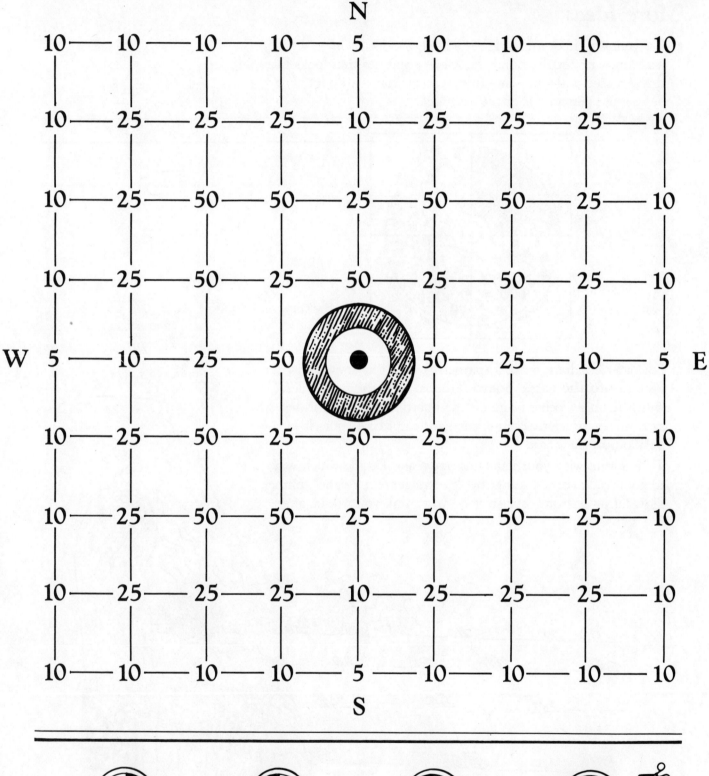

Chapter VII

TIME and MONEY

TOOL KIT

Clocks of all kinds (digital and analog, or old fashioned round), including some old ones that don't work

A watch or clock with a second hand

Calendars

Paper

Pens, pencils, and crayons

Scratch paper

Tape

Play or real money

Catalogs

Newspapers

Menus

Calculators

TIME AND MONEY

The topics of **Time** and **Money** are usually included in the curriculum as part of measurement. For this book, however, we have moved them into a separate chapter because they are of such importance to children's mathematics outside of school. Time and money are part of our lives, and everyday experiences will provide far more learning than can happen in classrooms. Every trip to the grocery store, every event that requires noticing the time is an opportunity to help children understand these concepts.

Very young children often have ideas about time that are hard for adults to understand. For example, a child may believe that anyone larger than another person is older, or that all adults are the same age. The idea of elapsed time, or how long certain things will take, is usually not well developed in children until they are nine or ten years of age. Just try explaining the concept of "wait" to a three-year-old! A child may believe that the minute hand takes longer to go once around the clock than the hour hand takes to move one hour ahead, because the minute hand went farther.

The important thing to remember is that children should not be rushed in learning about time. If your first-grader cannot tell time to the minute, don't frustrate yourself and him or her by insisting on trying to teach it. Be relaxed, don't fuss, and time will take care of itself. Do you know any adults who can't tell time?

There are some "skills," however, that you and your child can develop and have fun with. While you time them, have the rest of your family close their eyes until they think fifteen seconds have gone by. Then increase the time to 30 or 60 seconds. Try again and see whether they can come closer. Or check the clock and then work together on a task until you think five minutes have gone by, then check the clock again. Accurate estimation of how much time has passed is a useful skill, one many people never develop.

Money includes a wealth of educational possibilities. It is usually the first place for use of decimals, and continues to be a source of understanding as children work with more complicated decimals. For example, if a sixth grader cannot remember where to put the decimal in the problem 32.61×5, the problem can be worked out using pennies, dimes, and dollars.

Money is also an easy source of natural problems for you and your children to work out together, and usually provides high motivation and ease of understanding that is not available in more contrived arithmetical situations.

As with the other activities you will be doing, we recommend that you try to enjoy rather than endure. The trip to the grocery store should not become a drill and practice session, but a chance to talk together about the numbers and what they mean.

Time Activities

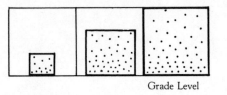

TOOLS

Clocks of all kinds

Why

To have a variety of experiences involving clocks and time

How

Double Clocks

☐ Help your young child learn how to tell time by putting a digital clock and an old-fashioned clock side by side. Leave them together, and have the child look at the times when he or she is going to eat, go out to play, or do some other thing.

Drawing Clocks

☐ Have your child draw clocks that show the times she or he gets up, goes to school, eats lunch, and goes to bed. Write the "digital" time beside each clock.

Time Line

☐ Make a time line of pictures of daily events. Have your child draw the pictures or cut them out of magazines.

☐ Make a similar time line for the different seasons in a year, or the holidays celebrated by your family each year.

What Time Will It Be?

☐ Get an old clock and move the hands to help you solve these problems:

If it is 6:00 now, what time will it be:

2 hours from now?

5½ hours from now?

8 hours from now?

12 hours from now?

What Time Was It?

☐ Use the old clock hands to help you solve these problems:

If it is 7:00, what time was it

1½ hours ago?

9 hours ago?

45 minutes ago?

2 hours and 20 minutes ago?

Just a Minute

- For these, use a watch with a second hand:
- Close your eyes and have another person watch the stop-watch. Open your eyes when you think a minute has gone by. How close did you come? Can you practice and come closer? Try 30 seconds.
- How long can you stand on one foot? (No holding on with hands, no touching with the other toe!)
- How long does it take you to brush your teeth? (In this one, longer might be better, you know.)
- How many times can you touch your knee and then your shoulder in fifteen seconds? Try this with both hands, to see whether there is a difference.
- How many times can you snap your fingers in fifteen seconds?
- How many times can you blink your eyes in 30 seconds?
- How long does it take you to get dressed in the morning?
- How long does it take you to tie your shoes?
- How many times can you write your name in one minute?
- How long does the toaster take to toast a piece of bread?

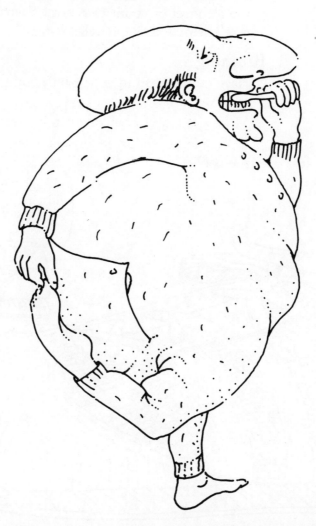

How Long?

☐ If it is 2:00 p.m. now, how long is it until 8:00 p.m.?

☐ How long is it until 11:30 p.m.?

☐ Make up more problems, and use the clock to work out the answers.

Mile Time

☐ When you are riding in a car, have your family try to guess when you have gone one mile, five miles, or ten miles. Check the odometer to see how close they came. Have them try again to see if they can come closer. Try this for one kilometer, five kilometers, or ten kilometers if possible.

Walking

☐ Find out how many miles your family can walk in an hour. Time yourselves by walking once around a quarter-mile track and multiplying by four. Or time yourselves walking around the track four times.

Contest

☐ Have a time race with two people to see if it takes longer to say the alphabet or count to twenty. Switch roles and re-time, in case one of you simply talks faster.

Running

☐ Have each person in your family run a quarter mile every day for a week.

☐ Make a graph of how long it takes each time.

☐ If possible, do the same for swimming back and forth across the pool.

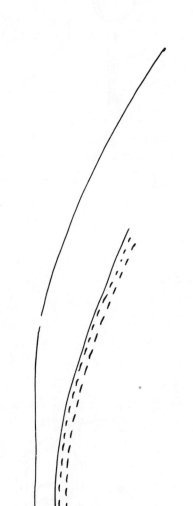

Time Zones

☐ Pick a city across the country. Find out what time it is there when it is 2:00 p.m. at your house.

☐ Find out what time it is in Oslo, Norway when it is 12:00 p.m. at your house.

☐ Find out what time it is in Hong Kong when it is 12:00 p.m. at your house.

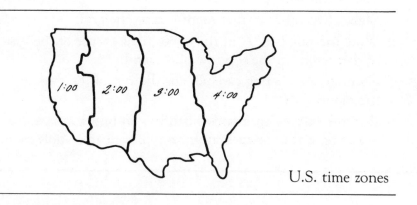

U.S. time zones

International Dateline

☐ Find out about the International Dateline.

☐ If it takes 18 hours to fly from Sydney, Australia to San Francisco, California and you leave San Francisco at 6:00 p.m. on Tuesday, what time and day will you arrive in Sydney?

☐ If you leave Sydney at 6:00 p.m. on Saturday, what time and day will you arrive in San Francisco?

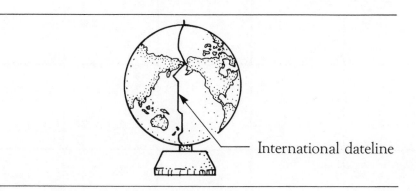

International dateline

Shadow Time

☐ Find a sharp-pointed shadow on a sunny day, such as that of the top of a pole or the peak of the house. Draw a ring around the point, six inches in diameter. Estimate **how long** it will take the shadow point to reach the ring. See whether you can also predict **where** the shadow point will touch the ring.

Making a Calendar

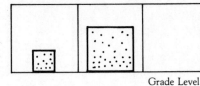

Grade Level

TOOLS

Calendar grid
Pencil or pen
Crayons or pens

Why

To help children learn about the calendar

How

☐ Make a copy of the calendar grid below.

☐ Help your child find out what day of the week the first of next month will be.

☐ Make a calendar for that month using the grid.

☐ Write the number "1" in the upper right corner of the first day of the month in the top row of the grid.

☐ Continue numbering carefully for the rest of the days of the month.

☐ Fill in holidays and friends' birthdays with pictures, and draw a picture at the top or bottom to represent the whole month.

Sunday	Monday	Tuesday	Wednesday	Thursday	Friday	Saturday

Calendar Patterns

Why

To provide opportunities to explore numerical patterns that occur on calendars.

> ► *Observation of patterns can help increase understanding and usefulness of the documents in which the patterns occur, such as calendars. Knowing, for instance, that there can be four or five Tuesdays in a given month, but never three or six, can be helpful in setting up a series of business appointments.* ◄

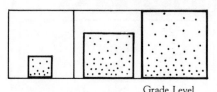

Grade Level

TOOLS

Calendar
Pencil and paper

How

☐ Find a calendar for next month—or use one you and your children have made.

☐ Count the number of Sundays, Mondays, and so on, there are in the month.

☐ Do some days occur more times than others? Which ones?

☐ Make a list of the Tuesday dates, the Wednesday dates, and Saturdays. Can you find the patterns in these lists?

☐ If the first Tuesday is on the 5th, how can you find the date of the third Tuesday without looking at the calendar?

☐ Make a list of the dates for each pair of Fridays and Mondays. Do you find any patterns in this list?

☐ If Friday is the 15th, what date will the next Monday be?

SUN	MON	TUES	WED	THUR	FRI	SAT
					1	2
3	4	5	6	7	8	9
10	11	12	13	14	15	
	?					

☐ Look at a different month and compare the patterns. Are they the same?

☐ Look for dates in different months that occur on the same day of the week.

☐ Which months start on the same day of the week? Would this be different in leap year?

□ Put a rectangle around three consecutive dates on your calendar.

SUN	MON	TUES	WED	THUR	FRI	SAT
				1	2	3
4	5	6	7	8	9	10
11	12	13	14	15	16	17
18	19	20	21	22	23	24
25	26	27	28	29	30	

□ Find their sum and compare it to three times the middle number.
□ Try the same with a different three consecutive numbers.
□ Try the same with five consecutive numbers, comparing the sum to five times the middle number.

SUN	MON	TUES	WED	THUR	FRI	SAT
				1	2	3
4	5	6	7	8	9	10
11	12	13	14	15	16	17
18	19	20	21	22	23	24
25	26	27	28	29	30	

□ Put a rectangle around a 3×3 square with nine numbers in it on your calendar.

□ Compare the sum of these numbers with nine times the center number.
□ Try the same for other 3×3 squares. Find the average of the numbers in a 3×3 square (to find the average, add all the numbers and divide by the number of numbers, or in this case, nine).

SUN	MON	TUES	WED	THUR	FRI	SAT
	1	2	3	4	5	6
7	8	9	10	11	12	13
14	15	16	17	18	19	20
21	22	23	24	25	26	27
28	29	30	31			

Money Activities

Why

To develop familiarity with coins, their relative values, and uses of money

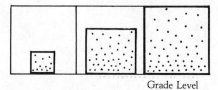

TOOLS

Money
Paper and pencil
Place value board
  (see page 178)
Catalogs

How

Pictures

□ Draw pictures of the front and back of all the coins.

Writing Money

□ Set out an amount of money, say $1.32, on a place value board, using only cents, dimes, dollars, and ten dollar bills. Then record the amounts of each coin on paper, being careful to write each amount in the correct place.

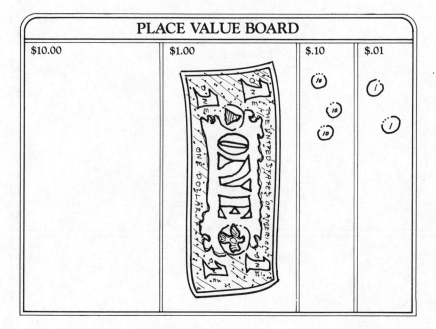

PLACE VALUE BOARD

$10.00	$1.00	$.10	$.01

□ Repeat this procedure many times, being sure to include some instances where there are no dimes or no dollars, such as $1.06, $10.45, or $50.07.

▶ *The zero place-holders are important to learn in writing money amounts.* ◀

PLACE VALUE BOARD

$10.00	$1.00	$.10	$.01

Catalogs

☐ Use any catalog that has items your children will find interesting.

☐ Pretend that you have $50 to spend together on gifts for your family, picking items from the catalog.

☐ Select the gifts so that the total will be as close as possible to $50.

☐ With older students, include sales tax in the calculations.

Class Party

☐ You have a budget of $25 for a class party.

☐ Prepare a shopping list that is within the budget.

☐ Check the prices in a newspaper ad or at a local grocery store.

Grocery Store

☐ Have your child accompany you to the grocery store, carrying a calculator.

☐ As each item is selected and put into your basket, have him or her round the price to the nearest dollar, and enter the dollar amount into the calculator.

☐ When you are ready to go to the checkout counter, ask your child for the expected total, then see how close that is to the actual amount you pay.

Comparison Shopping

☐ When you are in the grocery store, have your child compare two items to see which is more expensive.

☐ For example, is the larger package of cereal really a better buy than the smaller size?

□ To find this information, you will usually have to divide the price by the number of ounces in the package, and compare the **per ounce price** of the two items.

□ If possible, make the same comparisons in different stores. Are some stores more expensive than others?

Percents

□ Have your child find newspaper advertisements that indicate "20% off," or "marked down 50%."

Use a calculator to calculate the amount of price reduction and the final cost of these items.

□ Calculate the sales tax for several amounts of money, say $5, $10, $15.

Make a graph of the sales tax amounts up to $25.

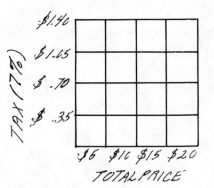

Menus

□ Gather together several menus from restaurants.

□ Have your children plan meals and find out how much the meals will cost, for themselves or for your whole family.

Pocket Survey

□ Take a survey of a number of friends, asking them how many of each kind of coin they have in their pockets or purses.

□ Graph the results.

Foreign Money

□ In the newspaper, usually near the financial section, there is often a list of the current value of foreign coins. An encyclopedia will also include information about foreign coins, or a bank might be willing to give information.

□ Choose a particular country, and practice making change in the money of that country, checking the value of the coins in relation to American money.

Chapter VIII

GEOMETRY and SPATIAL THINKING

TOOL KIT

Paper

Pencils, pens, and crayons

Watercolor paint

Scissors

Paste

Old magazines

Scraps of cloth or wallpaper

Cardboard or heavy paper

Graph paper
  (see page 79-82)

Tiles or paper squares

3″×5″ cards

Game markers

GEOMETRY AND SPATIAL THINKING

Geometry is the part of mathematics that includes information about shapes and space. Until fairly recently, the study of geometry was not introduced until the high school level, and then in a very formal way, with rules and proofs and strict definitions. With the introduction some years ago of "New Math," geometry instruction was moved down into the elementary grades, still usually taught in a relatively formal manner. It was soon found, however, that this kind of instruction led only to more memorization of words and ideas that did not make much sense to elementary children.

In the current curriculum, most educators feel that elementary geometry should be informal, allowing for exploration of ideas rather than memorization of terminology. Providing a background of intuitive understanding of the concepts can be very helpful for students who enter high school geometry classes.

For example, a student who has cut apart an isosceles triangle (one with two congruent or equal sides) and put the pieces together to make a rectangle has a better chance of understanding why the formula for the area of a triangle is 1/2 of its base times its height.

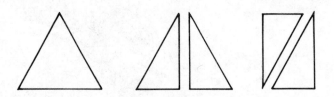

Doing geometry often requires the ability to visualize relationships of objects in space. Early experiences using real objects will help develop these skills. Spatial visualization skills are important both in math classes and in everyday life. Reading and sketching maps, giving and following directions, understanding diagrams and illustrations in putting together toys, furniture, and other household items, are examples where spatial thinking is required.

The essential ideas of symmetry and proportion are woven into our lives—in architecture, clothing, commercial designs, science, art, recreation, natural phenomena, to name only a few instances. You and your children can find many figures and patterns that represent geometry. Just as children need many concrete experiences before they are ready to learn abstract numbers, so do they need many concrete experiences with geometric shapes.

Another important part of geometry is understanding the coordinate system. Algebraic information is represented graphically by plotting points on a coordinate grid. Some types of equations will result in a straight line, while others will make a circle, or a parabola, or other shapes. Being familiar with the coordinate grid can be a real advantage to students when they take formal geometry and algebra courses.

The activities in this section provide experiences in moving and visualizing geometric shapes, in recognizing spatial patterns, representing movements on a two-dimensional plane, and using spatial patterns to reinforce number facts. Practicing these skills at home will strengthen children's confidence in tackling geometry.

Simple Symmetries

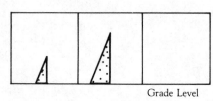

Grade Level

TOOLS

Paper (some colored)
Paint
Scissors
Crayons
Paste
Old magazines
Cloth or wallpaper scraps

Why

To help young children develop an understanding of bilateral symmetry and a sense of geometric patterns

▸ *Bilateral symmetry shows the same pattern arranged symmetrically on two sides of a center line or axis.* ◂

How

Help your child follow the directions for each design.

Blob Pictures

☐ Fold a piece of paper in half.
☐ Open it up and drop a blob of paint on one side.
☐ Fold it in half again and press.
☐ Open it up to see the design that has been made.

Cut Outs

☐ Fold a paper in half and cut a shape out along the fold.
☐ Guess what the shape will look like when the paper is opened.
☐ Paste the paper on another sheet to display the design.

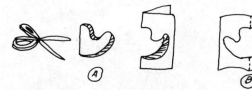

Flip Flop

☐ Cut several shapes out of the edge of a paper.
☐ Paste the paper onto another piece that is twice as long.
☐ Then take the pieces cut out, fit them into their original holes, and flip flop them to the other side so that they make a reverse design. Paste them into place.

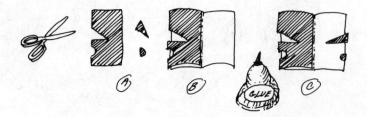

Snowflake

- ☐ Start with a square.
- ☐ Fold a square in half.
- ☐ In half again.
- ☐ In half again, on the diagonal.
- ☐ Cut out designs.
- ☐ Open the paper and find a snowflake pattern.

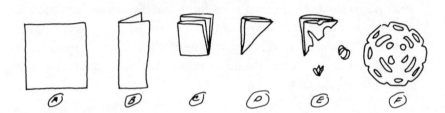

Symmetrical Pictures

- ☐ Fold a piece of paper in half.
- ☐ Color the same design on both sides so that it is symmetrical.
- ☐ To check, open the picture slowly to see if you see the same thing on both sides at the same time.
- ☐ If you start drawing from the outside and work in, it will be easier.
- ☐ Make your first picture simple.

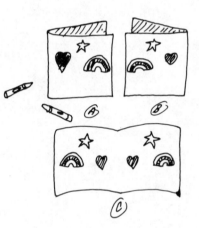

Alphabet Symmetry

- ☐ Think of the capital letters of the alphabet.
- ☐ Pick one, say C, and write it backwards. Does it look the same as it did originally?
- ☐ Pick another, say D, and write it backwards. Does it look the same?
- ☐ What happens with A? A is different, because it is symmetrical —both sides of the letter look the same.
- ☐ Go through the alphabet, making a list of the letters that are symmetrical (look the same on both sides) and those that are unsymmetrical (look different when written backward.)

Household Symmetry

- ☐ Look around the house and find objects that are symmetrical. Draw a picture of each one.

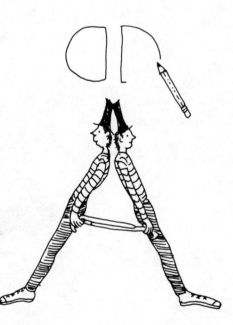

Complete the Symmetry

☐ Cut out pictures of designs from magazines or cut up scraps of wallpaper or cloth.

☐ Paste each picture in the center of a large piece of paper.

☐ With crayons or pencil, continue the design of the picture or cloth, until the paper is filled, or the symmetry is completed.

► *This activity is especially interesting to watch as young cildren draw their designs. Many children seem to be looking at the fine details of the picture, while others are seeing the larger whole. Often, children will see things that adults may miss completely. There is no right or wrong way, and children should be allowed to make their own interpretations of the designs.* ◄

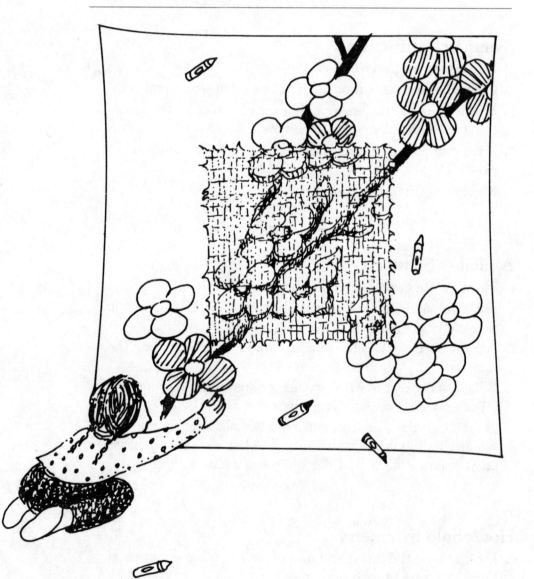

Create a Puzzle

Why

To explore the attributes of geometric shapes by building and solving a sequenced series of puzzles

How

☐ Start with a square or any other shape you find pleasing.
☐ Make one straight cut in any direction. For example:

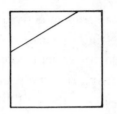

☐ Make a second cut. For example:

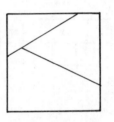

☐ Fit the three pieces together to make sure you can solve this puzzle.
☐ Make any third cut. For example:

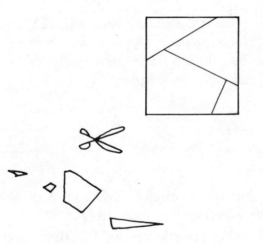

☐ Practice with the four pieces, then give your puzzle to a friend to solve.
☐ Special Note: If you wish to make your puzzle a little easier to solve, color the backs of the pieces differently from the front.

Grade Level

TOOLS

Scissors
Cardboard or
Heavy paper

Pentasquare Activities

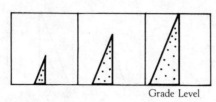

Grade Level

TOOLS

Pencil

2 cm graph paper
  (see page 80)

Scissors

Tiles or paper squares

Why

To develop spatial visualization skills

▶ *This activity will help students see* **congruent** *shapes and use a systematic approach for identifying characteristics of shapes. Two figures are* **congruent** *if they are exactly the same size and shape.* ◀

How

☐ This is a series of activities using shapes made of five squares.

☐ Use the tiles or paper squares at first to explore different arrangements, then record the results on graph paper.

☐ In arranging the five squares, the rule is that each of the squares must share a full side with at least one other square, and that wherever the squares touch, it must be with full sides touching.

 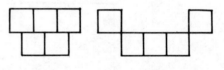

This is a pentasquare.   These are not pentasquares.

☐ Make as many pentasquares as you can, recording each on graph paper.

▶ *For this activity, we will consider pentasquares that are* **congruent** *as the same. If you can move, flip, or rotate one shape so that it fits exactly on another, the two shapes are congruent. These pentasquares are* **not** *different—they are congruent.* ◀

☐ Study the pentasquares carefully. Mark an X on those that you think can be folded up into an open box.

☐ Cut out all of the pentasquares and fold them to check.

☐ Sort out the pentasquares into those that will make a box and those that will not make a box. Look at the group that will not make a box and talk together about what you see and why you think those shapes didn't work.

More Ideas

☐ For younger children, try to find all the different arrangements that can be made with three squares, then four squares. None of these will fold into a box, but they can be sorted into those that have symmetry and those that do not. Cut out each shape and see if it can be folded in half to show that it is symmetrical. Not all of these shapes will work, but most of them will.

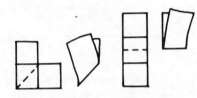

☐ For older children look for all of the shapes that have six squares. Some of these will make a cube.

If your children are interested in the project, try the shapes that have seven squares. There are a **lot** of them, so this activity will take extra time.

Cut-A-Card

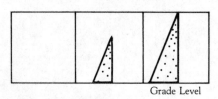

Grade Level

TOOLS

Cut-A-Card puzzle
Practice paper, 3″×5″
3″×5″ cards
Scissors

Why

To practice visualizing an object in space, observing the effects of flipping or rotating the object

▶ *Engineers and others who work in construction or design need to be able to visualize objects and their structure in many ways.* ◀

How

☐ Make a Cut-A-Card puzzle by cutting and folding a 3″×5″ card as indicated in the sketch.

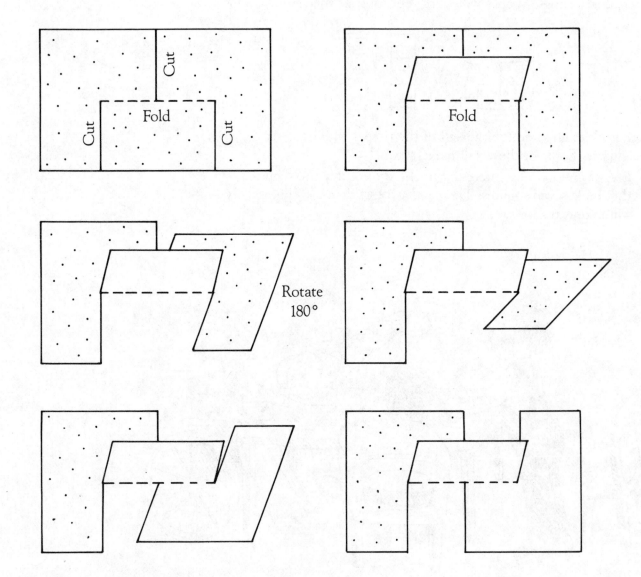

☐ Tape this object to a larger card or poster.

☐ Write the directions for the puzzle on your poster:

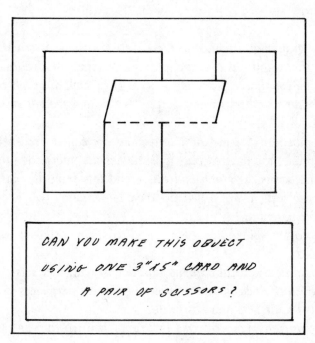

CAN YOU MAKE THIS OBJECT
USING ONE 3"X5" CARD AND
A PAIR OF SCISSORS?

☐ Put the puzzle poster out for your family or class to try. Have them use the small slips of paper for practice. Try to refrain from giving hints. You may be surprised at who finds the puzzle easy and who finds it hard.

More Ideas

Collect other three-dimensional puzzles to try. Game stores often carry a variety of commercial puzzles such as the Soma Cube. **Games** magazine is also a good resource for puzzles.

Coordinates I

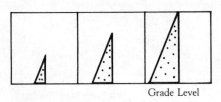

Grade Level

TOOLS

Graph paper
Pencils

Why

To learn the conventions of graphing points on the Cartesian plane

► *It is the graphing of points on the **Cartesian** or **coordinate plane** that allows us to draw graphs or pictures of algebraic equations. The coordinate plane is simply a grid, such as graph paper, with certain numbers naming certain **lines** and the **intersections** of the lines.*

► *The numbers that name the intersections are called **ordered pairs**. For example, the ordered pair (3,2) names the intersection of the third line across the graph and the second line from the bottom. These intersections are called **points**. The ordered pair is always shown in parentheses.*

► *The coordinate plane has a horizontal axis (across), which is numbered.*

► *There is also a vertical axis (up and down) with numbers.*

► *The point at which the horizontal and vertical axes cross is called the **origin**, and its name is (0,0).*

► *Mathematicians move **across** first, and then **up** to a particular point.*

► *Remember to move on the lines rather than the spaces.* ◄

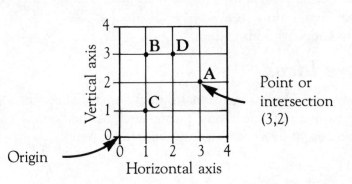

How

☐ Use the grid above to answer the series of questions, and to find the points.

☐ What letter is at (3,2)? This tells you to go over three lines and up two lines—and, of course, you found the letter A!

☐ What letter is at (2,3)? Go across two lines, and up three lines. Is this letter different from that you found with (3,2)? (Yes, it's D).

☐ What are the coordinates of the letter B? (1,3)

☐ Put a marker at (2,0).

☐ Use a larger piece of graph paper and make up your own problems.

FIRST QUADRANT Graph paper

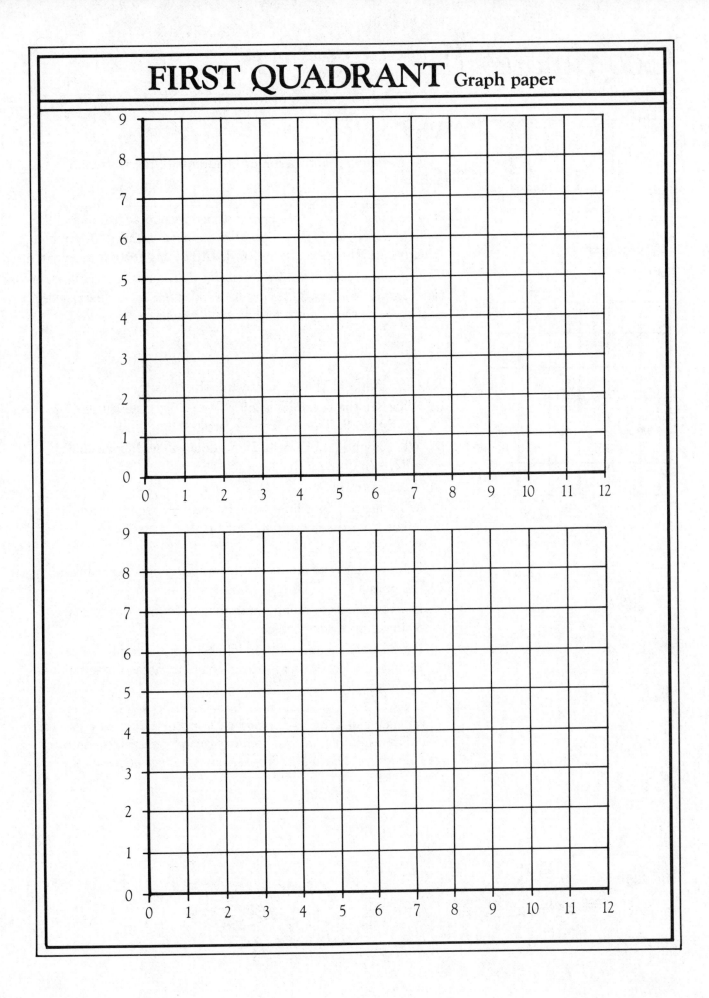

Coordinates II

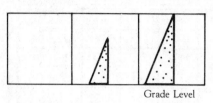

Grade Level

TOOLS

Graph paper
Pencil

Why

To become familiar with the four quadrants of the Cartesian graph

▶ *In Coordinates I, all of the named points were above and to the right of the origin, (0,0). This part of the plane is called the* **first quadrant**. *There are also* **second, third,** *and* **fourth** *quadrants. To indicate moving to the left or moving downward from (0,0), we use negative numbers, still going across according to the first number and up or down according to the second number.* ◀

How

☐ Use the graph to the left to answer the questions.

☐ What are the coordinates of F? We go 1 to the left and up 3 to get to F. The coordinates are $(-1,3)$.

☐ What about G? Go right 2 and down 2, with coordinates of $(2,-2)$.

☐ What point is at $(-2,-1)$? (E)

☐ What are the coordinates for H and I? How are these different from the coordinates for G?

☐ Put an X at $(0,-2)$.

☐ Point to different points on the graph, practicing naming the points.

☐ Use the sheets of graph paper that have four quadrants to make up more problems.

☐ Put all of the letters of the alphabet onto a graph. Write messages in code, using ordered pairs to name the letters you want.

▶ *Coordinate graphs are very important in mathematics, since they are the place where algebra and geometry come together. Understanding the Cartesian plane will help in advanced mathematics.* ◀

FOUR QUADRANTS Graph paper

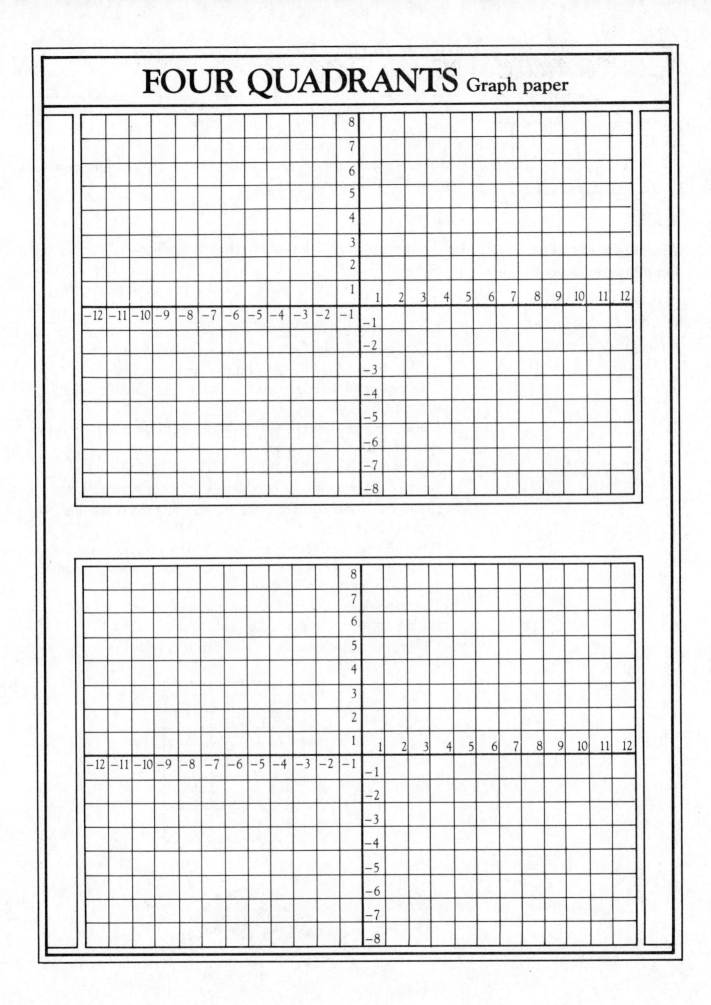

Coordinate Tic-Tac-Toe

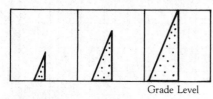

Grade Level

TOOLS

Graph paper (10×10)
Pencils or markers

A game for
2 players or 2 teams

Why

To reinforce the skills of naming coordinates and graphing points on the Cartesian plane

How

☐ The game is played almost like the old familiar Tic-Tac-Toe, except that:
  ☐ The X's and O's are put on the line intersections instead of in spaces.
  ☐ The board is larger—usually 10 by 10.
  ☐ The goal is to get four X's or four O's in a row.
  ☐ The places where the X's and O's are put must be given according to their ordered pair names (see Coordinates 1, page 192).

☐ Markers may be used instead of X's and O's, so that the board may be used over and over.

☐ First, decide on a leader. For the first game, this should be a parent or other adult. In later games, anybody can be leader.

☐ Number a 10×10 sheet of graph paper, using the coordinate system.

☐ Players (or teams if there are more than two players) take turns naming the points for the X and O. The points **must** be named by their ordered pair designations.

☐ The leader keeps a record on the grid (with pencil or with markers) of the points called by each team.

☐ The goal is to get four X's or four O'x in a row.

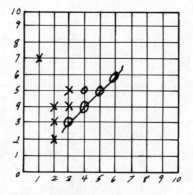

More Ideas

☐ Use a board that includes all four quadrants, so that some of the points will include negative numbers.

☐ Play with more than two players or teams. Four is a good number. Choose two more colors of markers, or two different symbols, such as A and B, in addition to the X and O. This game may take some unusual turns, and usually leads to cooperation between some teams or players.

Hurkle

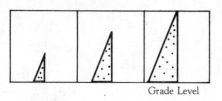

Grade Level

TOOLS

Hurkle paper
Pencil
Markers

A game for
2 or more players

Why

To practice naming points on a coordinate grid and using compass directions to find the hidden "Hurkle"

How

☐ Review the directions for naming coordinates (see page 194).

☐ Explain or review compass directions: North, South, East, West, Northeast, Southwest, etc.

☐ Choose a leader for the first game. Other players should have a turn leading later games.

☐ The leader decides on a point where the Hurkle is hiding and announces that a small, fuzzy, creature is hiding behind some point on the grid.

☐ The other players need to discover what the point is.

☐ Players take turns guessing coordinates, naming them by ordered pairs, such as (6,8).

☐ The leader responds to each guess with a clue, telling the players what direction they need to go from their guess to find the Hurkle. For example, if the Hurkle is hiding at (6,8) and the guess is (2,4), the leader will say "Go northeast."

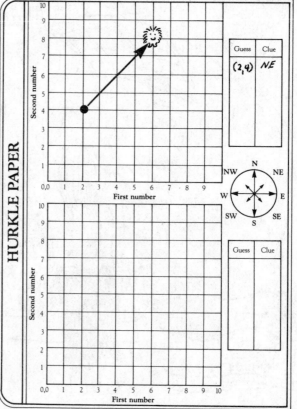

□ Players keep track of their guesses and clues.

□ The leader should mark the Hurkle's hiding place on a hidden sheet of Hurkle paper. After each guess, he or she should make a mark or place a finger on the guessed point and then give the direction players need to move to find the Hurkle. This helps avoid a common mistake of giving the opposite direction, or the direction from the Hurkle to the guess.

□ Be sure to talk about the best strategy for making guesses.

More Ideas

Play on a grid including all four quadrants.

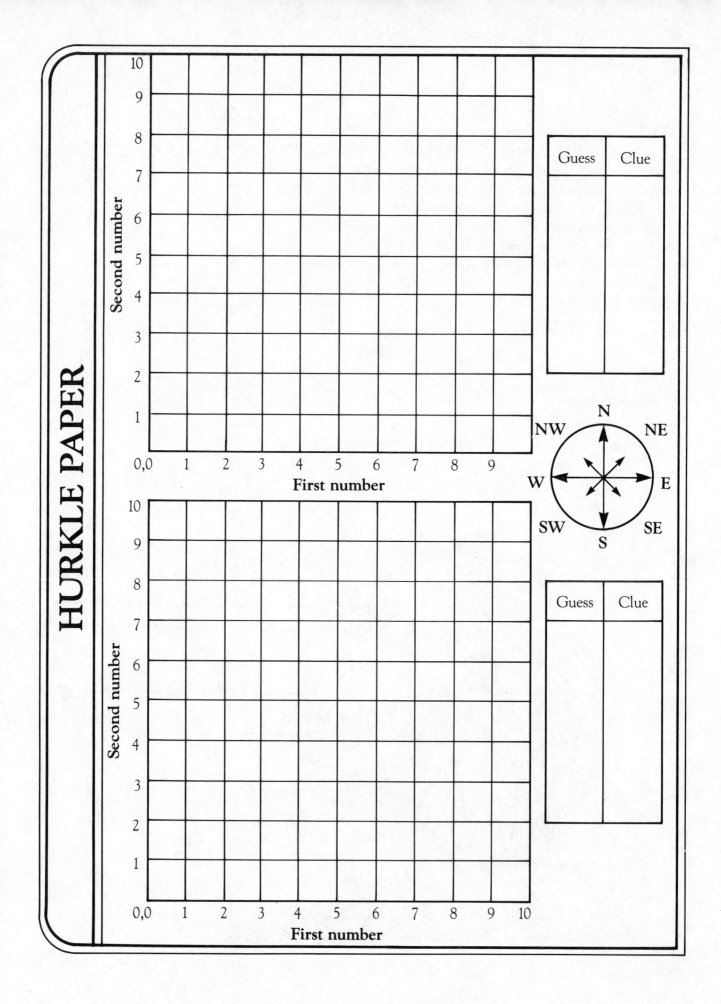

HURKLE PAPER

Chapter IX

PATTERNS
and
NUMBER
CHARTS

PATTERNS AND NUMBER CHARTS

TOOL KITS

Number charts
(see pages 203-205)

Markers (beans, small blocks, buttons, and so on)

Spinners

Crayons, pencils, pens

Familiarity with the patterns of number charts of all kinds is one of the keys to success in mathematics. A child who has had many experiences working or playing with charts will not only develop skills in interpreting and using charts but also will gain a deeper understanding of the attributes and relationships of the numbers and will be more likely to be able to understand the numbers as they are used later in fractions or in algebra, geometry, and calculus.

In this chapter, we introduce a group of activities that are based on grids of counting numbers or multiplication tables. The activities range from finding all of the numbers that include the numeral "2" to finding prime numbers or finding the patterns of palindromic numbers.

The first six activities use counting number charts in several versions. One style includes the numbers one through 100; a second version includes zero through 99; and a third, for use with younger children, includes one through 25. The activities can also be done with other grids of numbers, such as a seven by seven grid of the numbers one through 49. When you and your family have tried the activities given, explore other grids that depict different number relationships.

The last activity uses a multiplication table to help demonstrate the connections between different numbers and their multiples.

HUNDRED CHART

1	2	3	4	5	6	7	8	9	10
11	12	13	14	15	16	17	18	19	20
21	22	23	24	25	26	27	28	29	30
31	32	33	34	35	36	37	38	39	40
41	42	43	44	45	46	47	48	49	50
51	52	53	54	55	56	57	58	59	60
61	62	63	64	65	66	67	68	69	70
71	72	73	74	75	76	77	78	79	80
81	82	83	84	85	86	87	88	89	90
91	92	93	94	95	96	97	98	99	100

NINETY-NINE CHART

0	1	2	3	4	5	6	7	8	9
10	11	12	13	14	15	16	17	18	19
20	21	22	23	24	25	26	27	28	29
30	31	32	33	34	35	36	37	38	39
40	41	42	43	44	45	46	47	48	49
50	51	52	53	54	55	56	57	58	59
60	61	62	63	64	65	66	67	68	69
70	71	72	73	74	75	76	77	78	79
80	81	82	83	84	85	86	87	88	89
90	91	92	93	94	95	96	97	98	99

TWENTY-FIVE CHART

1	2	3	4	5
6	7	8	9	10
11	12	13	14	15
16	17	18	19	20
21	22	23	24	25

Cover Patterns

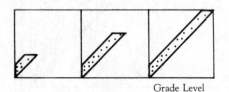

Grade Level

TOOLS

Hundred charts
Markers or beans

Why

To see visual patterns among the first hundred numbers

How

☐ Choose one of these rules, and cover all of the numbers on the chart that fit the rule. Usually it is better to take off the markers for one rule before starting another rule, but sometimes you may want to see how the rules overlap. Try these rules:

☐ numbers with a 2 in them

☐ numbers with a 4 in them

0	1	2	3	✕	5	6	7	8	9
10	11	12	13	✕	15	16	17	18	19
20	21	22	23	✕	25	26	27	28	29
30	31	32	33	✕	35	36	37	38	39
✕	✕	✕	✕	✕	✕	✕	✕	✕	✕
50	51	52	53	✕	55	56	57	58	59
60	61	62	63	✕	65	66	67	68	69
70	71	72	73	✕	75	76	77	78	79
80	81	82	83	✕	85	86	87	88	89
90	91	92	93	✕	95	96	97	98	99

☐ numbers with a 7 in them

☐ numbers with a 0 in them

☐ numbers with a 5 in the tens' place

☐ numbers with both digits the same

☐ numbers whose digits add to 9

For example, in the number 45, the digits 4 and 5 add to 9; or in the number 81, the digits 8 and 1 add to 9.

☐ numbers whose digits have a difference of 1

For example, in the number 45, there is a difference of 1 between the 4 and the 5; and in a 54, there is also a difference of 1 between the 5 and the 4.

☐ numbers that are multiples of 3

☐ numbers that are multiples of 5

☐ numbers that are evenly divisible by 6

☐ numbers that have a circle

☐ numbers that have a factor of 4

☐ Study the patterns that the different rules make. Mathematics **does** make sense when we see how it fits together.

□ Make some new cover pattern rules for your family to try.

1	2	3	4	5
6	7	8	9	10
11	12	13	14	15
16	17	18	19	20
21	22	23	24	25

1	2	3	4	5	6	7
8	9	10	11	12	13	14
15	16	17	18	19	20	21
22	23	24	25	26	27	28
29	30	31	32	33	34	35
36	37	38	39	40	41	42
43	44	45	46	47	48	49

1	2	3	4	5	6	7	8	9
10	11	12	13	14	15	16	17	18
19	20	21	22	23	24	25	26	27
28	29	30	31	32	33	34	35	36
37	38	39	40	41	42	43	44	45
46	47	48	49	50	51	52	53	54
55	56	57	58	59	60	61	62	63
64	65	66	67	68	69	70	71	72
73	74	75	76	77	78	79	80	81

Before or After

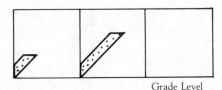

Grade Level

TOOLS

Hundred charts

**Markers or beans of
different colors**

**BEFORE/AFTER spinner
(see page 154
for directions)**

*A game for
2 players*

Why

To practice moving on a number line

How

☐ Make a spinner like the one shown.

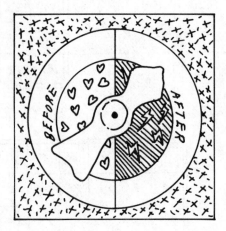

☐ Cover a hundred chart so just the first row shows, making
a number line from 1 to 10.

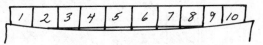

☐ Players take turns.
☐ On your turn, choose a number from 0 through 9, say 7.
☐ Then spin the BEFORE/AFTER Spinner.
☐ If you get a BEFORE, place one of your markers one number
before the number you chose. If you had chosen a 7, you
would put your marker on the 6.
☐ If you get an AFTER, place one of your markers one number
after the number you chose. If you had chosen a 7, you would
put your marker on the 8.
☐ If a marker is already on a number, you may not put another
marker on it. You may, however, choose a covered number
before you spin the spinner.
☐ Continue until all of the numbers are covered.
☐ The player with the most markers on the board wins.

More Ideas

☐ Play with a larger board, such as the numbers 0 through 19.
☐ Play with the rule that you must cover the number that is **two**
spaces before or after the chosen number.

More or Less

Why

To practice adding numbers and to become familiar with the hundred chart relationships in all directions—horizontally, vertically, and diagonally

How

□ Make a spinner like the one shown.

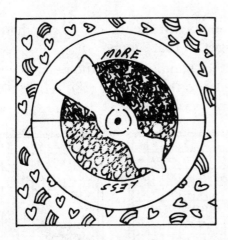

TOOLS

Hundred chart
Markers for each player
MORE/LESS spinner
   (see page 154
   for directions)

*A game for
2-4 players*

□ Use either a portion or the whole hundred chart.
□ Two to four players take turns spinning a MORE/LESS spinner.
□ The goal is to get four markers in a row—horizontally, vertically or diagonally.
□ At the beginning of the game, your group agrees on a **game number,** say 5.
□ On your turn, choose any number on the board, say 27, and spin.
□ If you get MORE, add 5 to 27 (or count on 5 from 27) and place your marker on 32.
□ If you get less, subtract 5 from 27 (or count back 5) and place your marker on 22.
□ Look for patterns that will help you get four markers in a row.

More Ideas

If you prefer, this game can be played cooperatively, with each person taking turns, trying together to make the row of four markers, or to make some other path across the chart.

Hundred Chart Operations

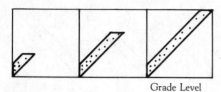

Grade Level

TOOLS

Hundred chart

Why

To gain understanding of place value by doing addition and subtraction problems on the hundred chart

> ► *This activity demonstrates graphically the meaning of place value, and the difference between the tens' and the ones' places.* ◄

How

☐ Point to any number in the top half of the chart, say a 25. Keep your finger on that number.

  ☐ With the other hand, point to the number you would go to if you added 3 to that number (28). Which direction did you go? (across to the right)

  ☐ Now point to the number you would go to if you **subtracted** 3 from the original number (22). Which direction did you go? (across to the left)

  ☐ Point to the number you would go to if you added 10 to the original number (35). Look for the shortest path.

☐ What about 20? 30? What pattern do you notice when you add tens to the number? (The answers are in the column directly below the original number, and the shortest path is straight down.)

☐ Now let's try adding a more difficult number to our 25. Let's add 34.

  ☐ First put your finger on the 25.

  ☐ Next, add the number in the tens' place, or 30. Moving straight down from the 25, say "10, 20, 30," moving one space down for each number. You should now be at 55.

  ☐ Then add the number in the ones' place, or the 4. Moving to the right from the 55, say "1, 2, 3, 4," moving one space to the right for each number. You should now be at 59.

  ☐ And there you are!! 25 + 34 = 59

- [ ] Try adding 25 + 37. You will reach the end of the row and have to go back to the first number of the row below. This is the same place in the problem where you would have to rename or carry a number.

- [ ] Subtracting seems to be much more difficult for most people, so be sure your child can do addition problems easily before trying to subtract. Subtracting requires moving in the opposite directions, or **up** for the tens and **left** for the ones.

  - [ ] Subtract 27 – 13. You will move up one ten or one row to 17, and then left three spaces to 14, and 27 – 13 = 14.

  - [ ] Try subtracting 25 from 82. Put your finger on 82, then move up, "10, 20," and then left "1", to the number 61— and here you will come to the end of the row.

  - [ ] Move across to the other side and up one row to the number 60, and continue on with "2, 3, 4, 5," to the 57.

- [ ] Practice with other numbers.

More Ideas

Use what you have learned here to play **More or Less** (page 209) with larger game numbers.

Hundred Chart Design

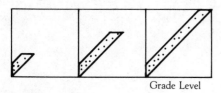

Grade Level

TOOLS

Hundred chart—8 copies
Crayons or marking pens

Why

To explore relationships of multiples and factors of the first 100 numbers

► *This activity provides a graphic representation of* **multiples** *and* **divisors** *of numbers, as well as* **common multiples and divisors** *of two or more numbers. At a glance you will be able to see that 6 is a multiple of both 2 and 3, that 30 is a multiple of 2, 3, and 5. By looking at the color pattern for each number, you can name its factors or divisors.*

Recognizing these patterns is important not only in memorizing multiplication and division facts, but later in reducing fractions and finding their common denominators, as well as in developing a good sense of numbers. The activity needs to be done slowly and carefully but is well worth the time. ◄

How

☐ Take a hundred chart and a crayon or marker of any color, say red.

☐ Write a large number 2 at the top of the page.

☐ Circle the 2 on the chart, then color in all the multiples of 2, or the numbers you would say if counting by twos—4, 6, 8,..., 100.

☐ Take another new chart and another crayon or marker, say brown.

☐ Write the number 3 at the top of the page.

3

1	2	3	4	5	6	7	8	9	10
11	12	13	14	15	16	17	18	19	20
21	22	23	24	25	26	27	28	29	30
31	32	33	34	35	36	37	38	39	40
41	42	43	44	45	46	47	48	49	50
51	52	53	54	55	56	57	58	59	60
61	62	63	64	65	66	67	68	69	70
71	72	73	74	75	76	77	78	79	80
81	82	83	84	85	86	87	88	89	90
91	92	93	94	95	96	97	98	99	100

☐ Circle the 3 on the chart, then color in all the multiples of 3, or 6, 9, 12,..., 99.

☐ Continue with new charts and different colors until you have a sheet for 2's, 3's, 5's, 7's, 11's, 13's, 17's and 19's.

☐ Put all of the charts out side by side, or hang them on the wall.

☐ Study the patterns you have made on your charts and have your family talk about them. What can you tell about a number when you look at it?

☐ Try to find all of the numbers that are not colored in on any of the charts.

▸ *These should be 23, 29, 31, 37, 41, 43, 47, 53, 59, 61, 67, 71, 73, 79, 83, 89, and 97. They are, along with the circled numbers, the prime numbers, meaning they have only themselves and one as divisors.* ◂

☐ Keep your charts up on a wall at home where you and your children can look for more patterns.

More Ideas

☐ Very young children may be helped by "counting on." For example, when they are finding the pattern of three, have them start with "1, 2, 3," putting a mark on the 3. Then, putting one finger on the 3, have them count again, "1, 2, 3," pointing at the 4, 5, and 6. Put another mark on the 6. Continue putting a finger on the last-marked number and counting on "1, 2, 3," and so on.

☐ Combine two or more charts.

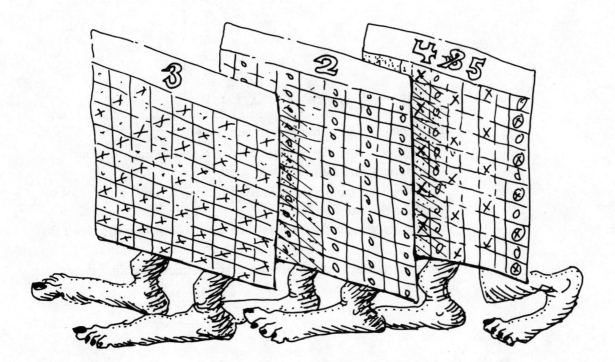

Palindromes

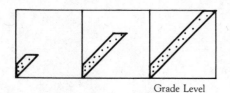

Grade Level

TOOLS

Palindrome chart
  (see page 216)

Crayons or marking pens
  of 6 different colors

Why

To develop accuracy in addition

▶ *This activity generates an interesting pattern on the hundred chart, and will give your family many hours of highly motivated addition practice. We suggest you and your children also practice using a calculator for at least part of the activity.* ◀

How

☐ A **palindrome** is a number that reads the same forward and backward such as 33, 868, 6006, or 52825.

☐ 423 is not a palindrome—BUT with a little addition 423 can be transformed into a palindrome:

$$
\begin{array}{r}
4\ 2\ 3 \\
+\ 3\ 2\ 4 \\
\hline
7\ 4\ 7
\end{array}
$$
A palindrome!

☐ We have written 423 backward and added it.

☐ We call 423 a 1-step palindrome, because we can turn it into a palindrome in 1 step.

☐ Some numbers take longer:

	5 9
	+ 9 5
Step 1	1 5 4
	4 5 1
Step 2	6 0 5
	5 0 6
Step 3	1 1 1 1

59 is a 3-step palindrome

☐ Choose some numbers of your own.

☐ Find out how many steps it takes to make each of your numbers into palindromes.

☐ With your family, explore the numbers from 0 to 99. Color all of the numbers that are already palindromes one color, the 1-step palindromes another, and so on. Use the chart on page 216.

☐ What patterns did you find after completing your Palindrome chart?

More Ideas

☐ Find palindromes for larger numbers.

☐ Look for words that are letter palindromes, such as NOON, or WOW.

☐ Write a phrase or sentence that is a palindrome.

☐ Read the story of ROBERT TREBOR by Marilyn Burns in **Good Times: Every Kid's Book of Things to Do.**

PALINDROME CHART

Choose a color for each:

☐ palindrome   ☐ 1 step palindrome   ☐ 2 step palindrome

☐ 3 step palindrome   ☐ 4 step palindrome

☐ 5 step palindrome   ☐ 6 step palindrome

0	1	2	3	4	5	6	7	8	9
10	11	12	13	14	15	16	17	18	19
20	21	22	23	24	25	26	27	28	29
30	31	32	33	34	35	36	37	38	39
40	41	42	43	44	45	46	47	48	49
50	51	52	53	54	55	56	57	58	59
60	61	62	63	64	65	66	67	68	69
70	71	72	73	74	75	76	77	78	79
80	81	82	83	84	85	86	87	88	89
90	91	92	93	94	95	96	97	98	99

Multiplication Designs

Why

To practice recognition of multiplication facts and to see connections between numbers and their multiples

▶ *Recognizing multiples quickly is important in addition and subtraction, reducing fractions, and long division, as well as in algebra.* ◀

How

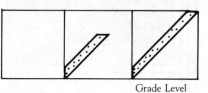

Grade Level

TOOLS

Several multiplication tables

Crayons or marking pens

☐ Pick a number from 2 to 12 and write it on top of your multiplication chart.

☐ Above your multiplication table, write all of the multiples of the number you chose. Continue your list until you get to 144, or as close as you can.

☐ For each multiple, find all the places it occurs on the table; color in those squares on your chart (see example below).

☐ Compare your design with one someone did for another number —or choose a different number from 2 to 12 and make a new design.

☐ Discuss the patterns made by different numbers. Why do 8 and 9 have more complex patterns than 7?

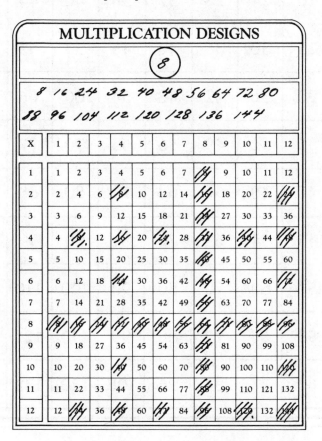

MULTIPLICATION DESIGNS

⑧

8 16 24 32 40 48 56 64 72 80
88 96 104 112 120 128 136 144

X	1	2	3	4	5	6	7	8	9	10	11	12
1	1	2	3	4	5	6	7	8	9	10	11	12
2	2	4	6	8	10	12	14	16	18	20	22	24
3	3	6	9	12	15	18	21	24	27	30	33	36
4	4	8	12	16	20	24	28	32	36	40	44	48
5	5	10	15	20	25	30	35	40	45	50	55	60
6	6	12	18	24	30	36	42	48	54	60	66	72
7	7	14	21	28	35	42	49	56	63	70	77	84
8	8	16	24	32	40	48	56	64	72	80	88	96
9	9	18	27	36	45	54	63	72	81	90	99	108
10	10	20	30	40	50	60	70	80	90	100	110	120
11	11	22	33	44	55	66	77	88	99	110	121	132
12	12	24	36	48	60	72	84	96	108	120	132	144

MULTIPLICATION DESIGNS

X	1	2	3	4	5	6	7	8	9	10	11	12
1	1	2	3	4	5	6	7	8	9	10	11	12
2	2	4	6	8	10	12	14	16	18	20	22	24
3	3	6	9	12	15	18	21	24	27	30	33	36
4	4	8	12	16	20	24	28	32	36	40	44	48
5	5	10	15	20	25	30	35	40	45	50	55	60
6	6	12	18	24	30	36	42	48	54	60	66	72
7	7	14	21	28	35	42	49	56	63	70	77	84
8	8	16	24	32	40	48	56	64	72	80	88	96
9	9	18	27	36	45	54	63	72	81	90	99	108
10	10	20	30	40	50	60	70	80	90	100	110	120
11	11	22	33	44	55	66	77	88	99	110	121	132
12	12	24	36	48	60	72	84	96	108	120	132	144

More Ideas

Color a design for two numbers on the same table. See **Designs from Mathematical Patterns** by Stanley Bezuszka, Margaret Kenney, and Linda Silvey for more number design ideas.

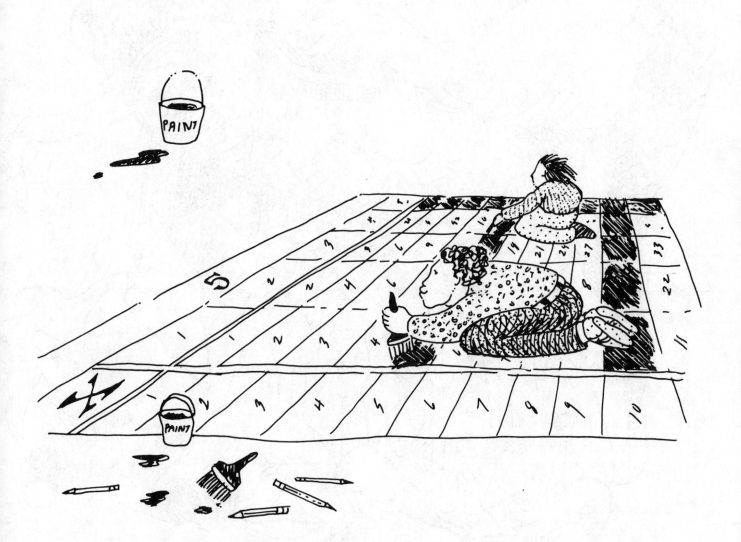

Chapter X

ESTIMATION, CALCULATORS and MICROCOMPUTERS

ESTIMATION, CALCULATORS AND MICROCOMPUTERS

The age of technology is changing mathematics educational needs. Students need to learn how to use these new tools, and they must be given the power of estimation.

TOOL KIT

Timer or watch with a second hand

Paper

Pencil or pen

5"×8" cards

Tape

Calculators

Game markers

Dice or spinner

ESTIMATION

Estimation skills are useful in every area of mathematics. In school, students first learn to approximate by rounding off numbers. They acquire techniques for estimating measurements of weight, length, area, volume, and of other quantities, such as the number of people at a picnic, or the number of peanuts in a package.

The calculator and the computer enable students to make faster and more astounding computational errors when they punch the wrong buttons or just don't think. Consequently, students need good estimation skills to decide on the reasonableness of an answer and to make decisions about how precise an answer should be.

People are often surprised to find that they rely so heavily on estimation to answer such questions as: What time should I leave home to get to the ballgame by noon? What size piece of cardboard do I need to practice my breakdance spin on? Should I take the freeway home during rush hour or would a side street be quicker? How long should I make the tail for my kite? Do I have enough cash to buy groceries?

A revealing exercise for parents and children in a FAMILY MATH class or at home is one developed by math educator Marilyn Burns:

☐ Write down ten ways in which you have used arithmetic in the past two weeks (outside of school).

☐ Then indicate whether each of the items required an exact answer or an approximate one.

☐ Next categorize the items according to whether they are done by paper and pencil, or mentally, or with a machine such as a calculator or cash register.

It seems that most adults rarely resort to paper and pencil to do the arithmetic that occurs in their daily life. They make estimates and use mental arithmetic—or they use a calculator or other machine.

USING ESTIMATION IN FAMILY MATH

The tricky part about teaching estimation is getting started. Even if you can convince people that estimating is useful, they will be hesitant to leave the security of calculations. You as a parent can be a model for your children by doing positive risk-taking activities such as Quick Questions (page 229), Calculator Paths (page 235), or Measure 15 (page 46) with your family.

Presented in a lighthearted, non-threatening manner, these activities will remind children and adults alike that the worst thing that can happen if anyone guesses an "incorrect" answer to a question in FAMILY MATH is *nothing*. No one takes out a red pen; there are no negative consequences for not having the "correct" response. In fact, making a mistake can be viewed as a great opportunity for learning and sharing information.

Akin to a willingness to risk being wrong is the notion of process over product. Both parents and children need to know that the processes they use or learn in problem solving are more valuable than any specific result. We find it easy to say all the appropriate words to our kids or to a FAMILY MATH class about the importance of process over product, and to appropriately praise explanations of strategies, but, too often, the most joyous, spontaneous praise is: "You got it! Wonderful!" which only praises the result. Our subsequent praise given for *how* the "correct" result was obtained often seems anticlimactic. We need to keep working on ways to show we really value process over product and to communicate that to our children and to adults.

Estimation is a lot like focusing your eyes on an object—things suddenly become much more clear and somehow more important. Attention is centered on the mathematical process rather than on the details of making a calculation, and the benefits can be impressive.

Once you are convinced of the need to be an estimator, and after you have encouraged your children or class to forget their inhibitions about making errors and to relax, the rest is easy. Just start estimating and asking your children or class to estimate at every possible opportunity.

At home you might ask your children to make the following type of estimates:

☐ If it is 4:30 now, what time will you get back from buying milk at the store?
☐ How many grapes in this bunch?
☐ How many toothpicks in the toothpick holder?
☐ How much will the groceries cost?
☐ What are the dimensions of the kitchen windows?
☐ How many cars will we pass on the way to the store?
☐ How many windows in that house?
☐ How many people will come to the soccer game?

Some examples for a FAMILY MATH class might be:

☐ I put the coffee on at 6:30; it's now 6:45; when do you think the red light will come on indicating it's ready?
☐ How many people are here tonight (estimate before you count)?
☐ How many squares do you think there are on this chart—more or less than 120? 150? 1000?
☐ I haven't estimated anything for over 30 minutes, who has something I can practice on?

The incentive and ability to make frequent informed guesses about the quantitative events in their daily lives will be a real mathematical gift to children.

CALCULATORS

The calculator is a powerful learning tool, and can help children learn all they are expected to master in the curriculum of today—problem-solving skills, geometry, probability, logical thinking, statistics, and so on.

We strongly encourage the development of skill in using calculators for all school-age children. These instruments are not just for gifted kids who already know their number facts and standard arithmetic algorithms, and not just for slower children who may never memorize their times tables or successfully complete a long-division problem. All children should have an opportunity to learn to use calculators efficiently, which requires a conceptual understanding of arithmetic. Hence, while calculator usage will shorten the time children will need to spend doing pages and pages of computation, it will, if anything, increase the need for children to learn how arithmetic works and what numbers really mean.

There is really no longer a need to be able to speedily crank out square roots of large numbers without calculators. More important is the need to know enough about roots and numbers to see quickly that $\sqrt{129,465}$ is between 300 and 400 and closer to 400 because $300 \times 300 = 90,000$, $400 \times 400 = 160,000$ and 129,465 is closer to 160,000 than to 90,000.

Changes in the marketplace and the labor force have created a critical shift in the questions children must now be able to answer about arithmetic. It is no longer sufficient just to know **how** to add, multiply, divide, or find percents. Equally important is knowing **when** it is appropriate to multiply or divide.

Until now, children have had to spend so much of their time doing sums and long-division drills that most learned only to be technicians rather than problem solvers. Test results show that our children are becoming better at paper-and-pencil computation, but are falling farther behind in reasoning skills. We need a shift in priorities to allow children to use machines for computation, using the time saved to learn better to do what the calculator cannot—to reason.

USING CALCULATORS IN FAMILY MATH

Calculators are also fun and fascinating objects of study. We have included a few activities that will make discovery of the capabilities and limitations of these electronic marvels interesting.

It is easier to talk about calculators if everyone has one. In a class situation, encourage the families to bring in their own calculators, or use a class set if one is available.

If you plan to purchase a calculator (or a set for class use), shop for calculators that automatically turn off when not in use, that are sturdy, and that have memory functions and percent keys. Such calculators are quite inexpensive; you can find them for as little as $5.00.

Calculators should be available at home and in FAMILY MATH classes in the same way that scratch paper, blocks, and beans are, to help solve problems.

Everyday Estimation

Why

To make estimation a conscious part of daily life

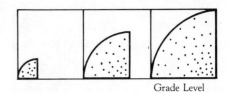

Grade Level

▶ *Estimation is a constant part of our everyday life. People shopping in the grocery store estimate whether they have enough cash to pay. Garage keepers estimate the cost of auto repairs. Contractors estimate the cost of buildings. Parents estimate the length of time before students get home from school. Estimation can also be a powerful tool in giving students control over more formal mathematics. When a person stops to estimate before solving a problem, the problem itself takes on more sense and becomes more manageable.* ◀

TOOLS

Timer or watch with second hand

Paper and pencil

How

The following list of questions is only a starter. As you and your family practice thinking about estimation, you will discover many additional possibilities. For each question, make a guess, write it down, and then find the answer to the question.

☐ How many times a day is the refrigerator door opened?

Tape a piece of paper to the front of the refrigerator, and make a tally mark each time it is opened. How close did you come?

☐ How many times do you chew each bite of food?

Ask another person to count for you, so that you can concentrate on eating! Try different kinds of food.

☐ How many pennies does the average adult have in his or her possession?

Ask at least ten people to give you an answer.

☐ How many pages are there in your dictionary?

If you have more than one dictionary, take an average.

☐ How many **1's** are there on a calendar?

Do you need a calendar to answer this one, or can you find the answer in some other way?

☐ How many books are there in the school library?

To get a start, count the books on several shelves, then count the number of shelves.

☐ How many windows are there in your house?

Count them.

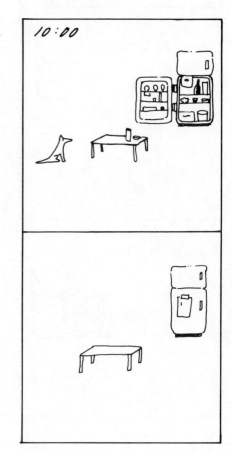

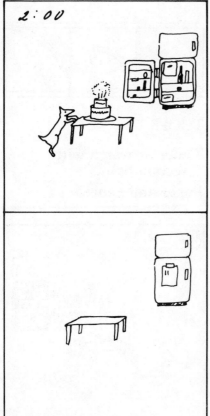

□ How many square feet are there in your living room?
Measure both directions, and multiply one measurement times the other. This will give you the area, or square feet.

□ How many letters of the alphabet are **not** included in the names of the days of the week?
Make a list of the letters and a list of the days. Compare the lists, crossing off the letters in both lists.

□ How many storage drawers are there in your entire house?
Guess you'd better count them!

□ How many grams of canned food does your family have in the cupboard?
Read the labels, make a list of the weights, and add it up.

□ How many steps is it from your front door to the back door or the back of the house?
Pace it off. Which is more accurate, big steps or heel-to-toe steps?

□ What is the longest word on the 75th page of your dictionary?
Look and count.

□ What is the difference between the longest and the shortest word on that page?
Look, count, and subtract.

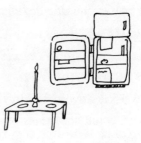

Why

To practice estimating

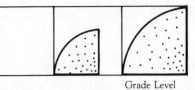

Grade Level

► *The ability to estimate has many facets. Willingness to make as informed a guess as possible is one component. Rounding off to make quick mental computations is another. Averaging and the ability to estimate averages are important skills*

► *NOTE: This activity needs a group of at least ten people. It might be done at a youth group meeting, or at a birthday party, or just on a rainy day when a lot of children are looking for something to do.* ◄

Preparation

☐ Write one of the questions from page 231 on each card. Prepare enough cards so that there is a question for each person. Some questions may be duplicates, or you may want to find some additional questions of your own.

☐ Make enough copies of the answer sheet so that each person can have one.

Directions

☐ Distribute the answer sheets, one to each person.

☐ Tape a question onto the **back** of each person, without letting them see what the question is.

☐ Explain to the group that they are going to **estimate** the answers to the questions on each other's backs.

☐ Each person will ask five other people to answer their question without reading it aloud. The answers are to be recorded on the answer sheets. Warn them that some will be easy to estimate and others more difficult.

☐ After getting five answers, each person writes down the range of his or her answers (the lowest response to the highest one.)

☐ They then estimate the average of the answers and compute the exact average. You may want to remind everyone how to find the average: sum the five answers and divide by five. Encourage those who like to use calculators to do so.

☐ When all have made their computations, have each person with the first question read his or her question and report the range and the average answer. Comment on extraordinary ranges. Allow the group to indicate whether they think the answer given is high, low, close, or far from the actual answer.

- ◻ Continue with other questions.
- ◻ After each question has been read and the estimates given, the leader should supply the best guess according to his or her sources.

► *Some of the questions have exact answers, and some do not. For some, even the answer is an estimate. You might want to talk about the use of information that is valuable, even though exact figures are not known. For example, a manufacturer must guess at how many tables to make; and that guess will be based on guesses about how many tables consumers will want and how many of this particular kind they will buy.* ◄

More Ideas

This activity may also be done by reading the questions aloud, one by one.

- ◻ Have people work in pairs to talk about their estimated answers.
- ◻ After all answers have been estimated, give the answers. (See page 258 for answers to the given questions.)

QUICK QUESTIONS

Estimate the answers to the following questions:

1. Approximately what is the sum of $3.09, $1.89, $0.49, and $2.51?

2. If 600 bottles are to be put in cases that hold 24 bottles, about how many cases will be needed?

3. Approximately how many seconds are there in a day?

4. You wish to dispose of one million dollars, so you give away $50 every hour; about how long will it take you to give it all away?

5. About how many times a day does the average American laugh?

6. About how many times a day does the average American blink?

7. Counting labels, signs, and everything else, how many commercial messages does the average American see in one day?

8. How many hot dogs does the average American eat in one year?

9. In 1983, over 45% of automobiles sold in the United States were purchased by women. What do you estimate to be the percent of microcomputers purchased by women in 1983?

10. How many 12-ounce cans or bottles of soda does the average American drink in a year?

11. How many miles long is the line drawn by the average pencil?

12. As of 1978 approximately how many boy scouts were there in Pakistan?

See page 258 for answers.

QUICK QUESTIONS ANSWER SHEET

The answers I received from other people were:

1. _____

2. _____

3. _____

4. _____

5. _____

Estimated average: _____

Average: _____

Range: from_____to_____

The answers I received from other people were:

1. _____

2. _____

3. _____

4. _____

5. _____

Estimated average: _____

Average: _____

Range: from_____to_____

The answers I received from other people were:

1. _____

2. _____

3. _____

4. _____

5. _____

Estimated average: _____

Average: _____

Range: from_____to_____

The answers I received from other people were:

1. _____

2. _____

3. _____

4. _____

5. _____

Estimated average: _____

Average: _____

Range: from_____to_____

Key Calculator Experiences
A prerequisite for calculator activities

Why

To learn the basic working of your calculator to prepare for years of happy computing and hours of fun with Aunt Bebe's Costly Calculator, Calculator Paths, Lost Numbers, Lost Rules, and more. Play around with your calculator—see what makes it tick.

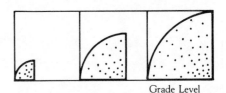

Grade Level

TOOLS

Calculator—at least one per family

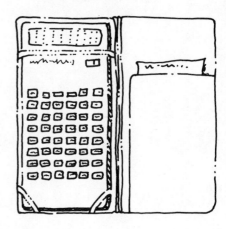

How

☐ **Turning on**—Look for ON/OFF keys or switches. Some calculators will turn off automatically if left on accidentally. Some solar-powered calculators do not have any OFF switch.

☐ **Clearing**—A CLEAR key will wipe out old information in your calculator. It is good to clear the calculator before starting a new calculation. To fix up a mistake made in the middle of a calculation, the "CLEAR ENTRY" CE key will erase just the last number put on the calculator display.

☐ **Computing** sums, products, differences, and quotients. Make up a set of problems for each person in your family to **estimate** and then **compute**.
For example,   for young members: and for middle school:

30 + 10	65 + 42
30 − 10	65 − 42
30 × 10	65 × 42
30 ÷ 10	65 ÷ 42

If a calculator has no equals sign = , it's not broken but uses a sophisticated nonalgebraic logic. Unless your family is quite familiar with calculators, it would be wise to start with a less fancy machine, one with an = key.

☐ **Constant** computing. It is possible to make most calculators remember a number and an operation to use over and over again. This "constant" feature works in a variety of ways depending on which calculator is used. A little determination and experimentation will yield the correct method for your machine.

   ☐ Try to count by 2's.
   ☐ Try 2+2====. That usually works. Some calculators prefer 2++2===. Figure out how yours works.
   ☐ Explore how to use a constant feature for multiplication, division, and subtraction.

☐ **Calculator** messages. Find out how your calculator will tell you that you tried to divide by zero, created a number bigger than it can handle, or put something into its special memory storage. Recognition of these messages will be helpful and can save you time later during long computations.

Calculator Paths

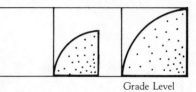

TOOLS

Game boards
  (see pages 236-239)
Markers of two colors
Calculator

A game for
2 players or teams

Why

To practice estimation and mental calculation

How

- ☐ Decide on the game board you wish to use. There is one for addition, addition and subtraction, multiplication, or division.
- ☐ The object of the game is to correctly choose, by estimating, numbers that will make a path from one side to the opposite side. Opposite sides are marked by the same symbol of stars or dots.
- ☐ Any path may be chosen, with as many turns as necessary to reach the other side.
- ☐ The directions on the boards indicate that **players** should take turns, but the game may be played with **teams** of two or more players working together.
- ☐ Players or teams use estimation to pick the two numbers they want to use on each turn.
- ☐ After announcing the choice of numbers to the other player or team, the sum, difference, product, or quotient is found using a calculator.
- ☐ The player or team places a marker on the answer if it is on the game board and not yet covered.

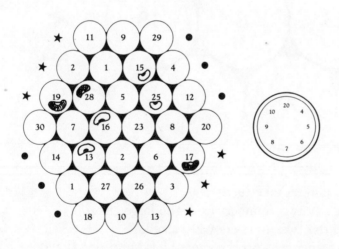

- ☐ If the answer is not available, the team cannot place a marker and it is the other team's turn. Some answers may not be on the board at all and others may already be covered.
- ☐ To win, a player or team must cover any path of answers connecting opposite sides as indicated on the game board.

CALCULATOR PATH Addition

★ Two players take turns.
★ Pick any two numbers from the large circle.
★ Add the two numbers you picked.
★ If the answer is on the game board and not already covered, cover it with one of your markers.
★ To win: Make any path with your markers that connects your two sides.
★ The first player goes from one star side to the other star side, and the second player goes from one dot side to the other dot side.

CALCULATOR PATH Subtraction and addition

- ★ Two players take turns.
- ★ Choose any two numbers from the large circle.
- ★ Add *or* subtract the two numbers.
- ★ If the answer is on the game board and not already covered, cover it with one of your markers.
- ★ The first player goes from one star side to the other star side, and the second player goes from one dot side to the other dot side.
- ★ To win: Make any path with your markers that connects your two sides.

CALCULATOR PATH Multiplication

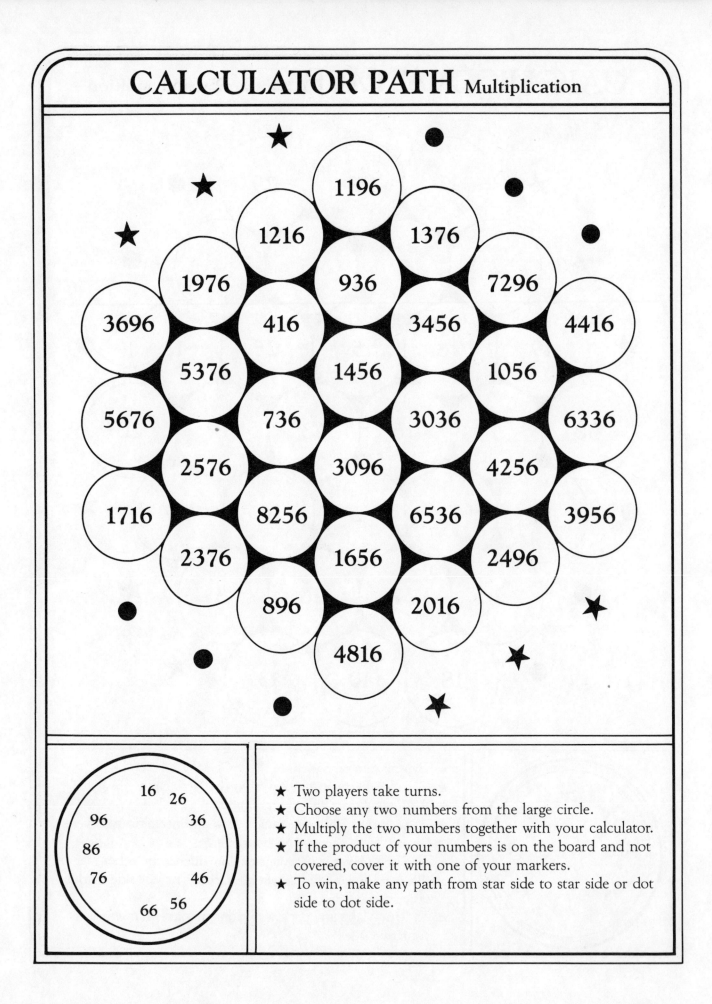

★ Two players take turns.
★ Choose any two numbers from the large circle.
★ Multiply the two numbers together with your calculator.
★ If the product of your numbers is on the board and not covered, cover it with one of your markers.
★ To win, make any path from star side to star side or dot side to dot side.

CALCULATOR PATH Division

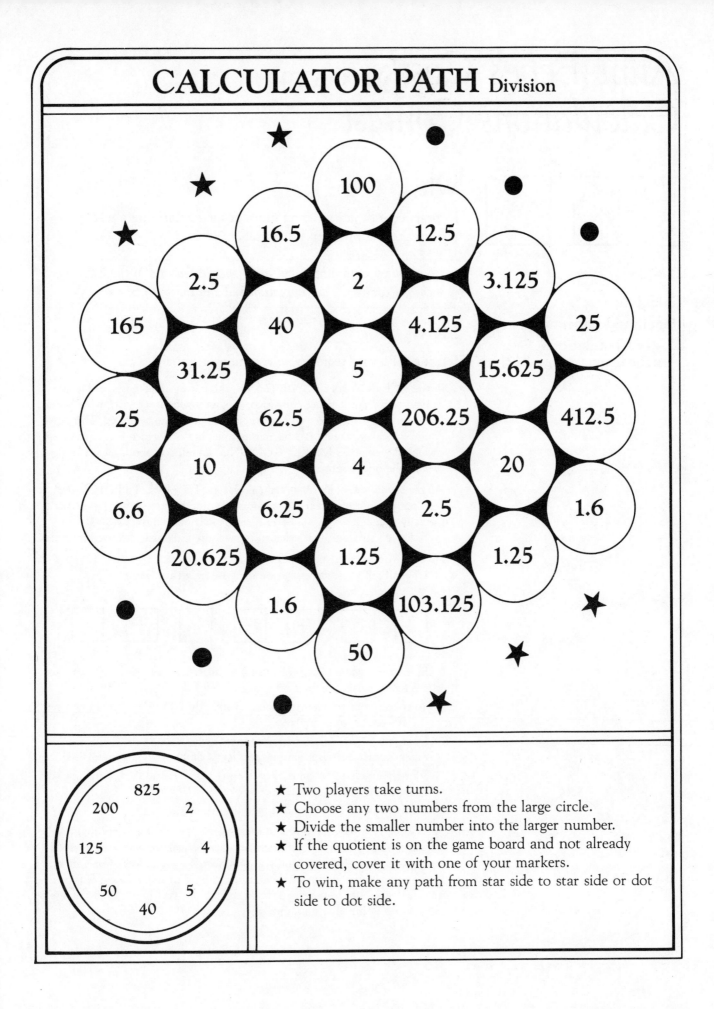

100
16.5
12.5
2.5
2
3.125
165
40
4.125
25
31.25
5
15.625
25
62.5
206.25
412.5
10
4
20
6.6
2.5
1.6
6.25
1.25
20.625
1.25
1.6
103.125
50

825
200   2
125   4
50   5
40

★ Two players take turns.
★ Choose any two numbers from the large circle.
★ Divide the smaller number into the larger number.
★ If the quotient is on the game board and not already covered, cover it with one of your markers.
★ To win, make any path from star side to star side or dot side to dot side.

Aunt Bebe's Costly Calculations Contest

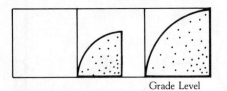

Grade Level

TOOLS

Calculators
Pencil and paper
Costly Calculations
  record sheet (page 242)

Why

To practice estimation using number facts, planning ahead, and breaking a problem into smaller pieces

► *Cooperation with others and using special features of the calculator such as "constant" are important skills in the business world.* ◄

How

Tell the following story to your family or your class:

In the wilds of Nedbury, North Dakota, there lives an eccentric old gentle-woman called Aunt Bebe. Aunt Bebe has a burning interest in many things, but potato pancake bake-offs and computing machines were two. The Great Potato Pancake Bake-off is a fascinating tale, but one which will have to wait until we discuss FAMILY COOKING. I'll tell you about a calculating contest she once sponsored.

All contestants were given special calculators. They looked a lot like ours with one interesting difference—most of the numbers were missing. In fact, the only number keys were 2 and 6. Aunt Bebe agreed to pay $1.00 for every key pressed on those calculators by contestants who followed her rules.

Here are the rules she gave them:
☐ Unless I tell you differently, you may use only these keys:

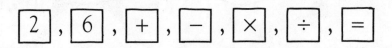

☐ In as many **ways** as possible, using **at most** ten key presses per way, get the following six numbers on your calculator display: 12, 30, 19, 13, 110, 6.2
☐ To collect your payment, you must write down your key strokes and get the answer on your display.
☐ I will pay $4 for each key stroke for each way you get the following numbers on your display, using **ten keys or less**: 0.16, 6.4, and 0.03
☐ When you use any memory keys to get 0.16, 6.4, and 0.03, using ten keys or less, I will pay only $0.25 per stroke.
☐ All solutions must be handed in within three days.

Aunt Bebe only sponsored one of these events; people got pretty rich and calculator-smart. Aunt Bebe's neighbor got so calculator smart she claimed that, given a calculator with only the number key 4 and lots of operation keys, she could get all the whole numbers from 0 to 100, and she did. Someone challenged her to use the digits in the current year to get all the numbers from 0 to 100. But that's another story, too.

How would you fare in Aunt Bebe's Costly Calculations Contest?

☐ Give each person a Costly Calculations Record Sheet.

☐ Go over the rules and examples.

☐ Try some numbers.

☐ Record your solutions for the first 6 numbers.

WAYS TO GET 12	WAYS TO GET 30	WAYS TO GET 19
1. 2×6	1. 6×6−6	1.
2.	2.	2.
3.	3.	3.
4.		

COSTLY CALCULATIONS
RECORD SHEET

★ Use only these keys:

$$2 \quad , \quad 6 \quad , \quad + \quad , \quad - \quad , \quad \times \quad , \quad \div \quad , \quad =$$

★ Get each of the following numbers on your calculator display: 12, 30, 19, 13, 110, 6.2
★ Make them appear on the display in as many ways as you can, **but** don't use more than **ten keystrokes** for any way.
★ If you earned $1.00 for each key stroke, how much money could you make?

★ For example,

$$2 \quad \times \quad 2 \quad \times \quad 6 \quad - \quad 2 \quad \div \quad 2 \quad =$$

is a $10 solution for 11.

Number	Keystrokes	Earnings
12		
30		
19		
13		
110		
6.2		

Tic Tac Toe–Rounding in a Row

Why

To practice estimation and rounding numbers

► *Estimation and rounding are essential skills for the calculator age. Playing this game encourages these skills and helps develop use of strategies.* ◄

How

► *This game is played in the same way as the classic game of Tic Tac Toe, except that the placement of markers is determined by adding (or multiplying) numbers together, and finding the* **number closest to that answer** *on the playing board.*
► *There are two gameboards, one for addition and one for multiplication. Both boards show numbers that are* **rounded to the nearest ten.** *The boards will not show exact answers, since one of the skills being practiced is rounding numbers.*
► *Use whichever board is appropriate for your family.* ◄

□ Players (or teams) take turns.
□ On each turn, the player chooses two numbers from the Addend Pool (or Factor Pool) and finds their sum (or product.) Calculators may be used to find these answers.
□ The player then finds on the board the rounded number that is **closest to** the sum or product and places a marker on the rounded number.
□ The first player to have four markers in a row, horizontally, vertically, or diagonally, wins the game.

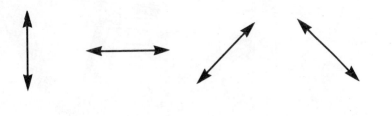

► *Rounding to the nearest ten is done as follows:*
► *If the number in the ones' place is four or less, the rounded number is the nearest multiple of ten that is less than the original number.*

For example, 762 is rounded to 760, or 76 tens.

► *If the number in the ones' place is five or more, the rounded number is the nearest multiple of ten that is more than the original number.*

For example, 768 is rounded to 770 or 77 tens.

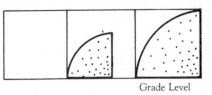

Grade Level

TOOLS

**Tic Tac Toe Game board—
Addition or multiplication
(see pages 245-246)
Markers of 2 colors**

*A game for
2 players or teams*

► *The same kind of rules apply in rounding to the nearest hundred, except that if the tens and ones together are 49 or less, the number is rounded down to the nearest hundred below the original number, while if the tens and ones are 50 or more, the number is rounded up to the nearest multiple of 100 that is more than the original number.*

For example, 369 becomes 400 or four hundreds and 128 becomes 100 or one hundred. ◄

CALCULATOR GAME BOARD Addition

★ Players take turns.
★ On each turn, the player chooses two numbers from the Addend Pool.
★ Find the sum of the two numbers. You may use a calculator.
★ Find the number on the board that is **closest to the sum,** and put your marker on that number.
★ The first player to have four in a row wins the game.
★ The numbers on the board are rounded to the nearest ten.

Addend pool

4	7	11	23	31	42
		49	62	70	

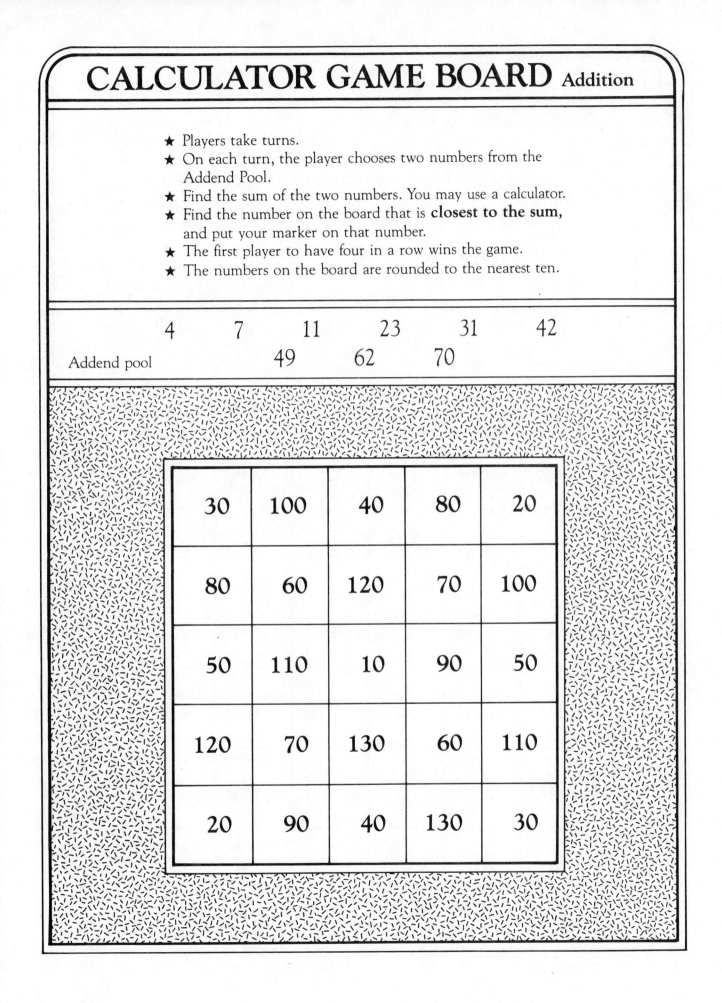

30	100	40	80	20
80	60	120	70	100
50	110	10	90	50
120	70	130	60	110
20	90	40	130	30

CALCULATOR GAME BOARD Multiplication

★ Players or teams take turns.
★ On each turn, the player chooses two numbers from the Factor Pool and finds their product. A calculator may be used.
★ The player then finds on the board the rounded number that is **closest to the product,** and places a marker on that number.
★ The first player to have four markers in a row, horizontally, vertically or diagonally, wins the game.

Factor pool

3	23	31	47	16	18
	17	59	13		

140	1460	180	940	390	1830
370	850	1080	210	50	1060
750	70	270	410	30	220
710	300	500	90	310	40
290	560	50	800	1360	770
50	610	1000	2770	230	400

The Lost Rules

Why

To practice pattern recognition and to develop understanding
of number relationships

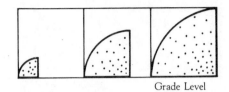
Grade Level

TOOLS

**Calculator with constant
feature**

▶ *A **function** is a special kind of mathematical rule that gives a unique
answer for each number. For example, the rule might be "times 4,"
so that if we think of a 3, the answer is 12, or if we think of a 10,
the answer is 40. The rules can be more complicated such as "times
2 plus 5" or "minus 6 divided by 10," although in this activity only
a single operation will be used.*

*Understanding of function rules and how they work is important
for students who intend to take higher mathematics courses such as
calculus. Playing games or working with functions is an excellent
way to give younger students practice in arithmetic.* ◀

How

☐ Before you introduce this activity to your family, practice entering functions into the calculator. An easy rule might be +3.
To enter this rule, press ⊞③⊟ . To hide the rule, enter any
other number, say ⑤ , then press ⊟ again. The display should
show 8. Try a few other numbers, remembering to press first
the number, then ⊟ .

☐ When you are sure you can remember how to hide numbers,
tell your family this story:

> One wintry day, when there was nothing to do, Aunt Bebe was rummag-
> ing in her closet for a game for the children to play, and ran across an
> old calculator. She exclaimed, "Well, now, I wondered what had become
> of this. Here's an old friend, but I had to stop using it because it kept
> losing the rules. In fact, I'll bet there are thousands of rules lost inside
> this calculator. Shall we try to find some?"
>
> Everybody agreed to help, so Aunt Bebe had each person take a turn
> giving a number. She would enter the number, then the equals sign,
> and write down the number that showed on the display.
>
> Here are the numbers the children guessed that first time, and the clue
> numbers from the display on Aunt Bebe's old calculator.

Entered number	Clues from Display
5	10
10	20
3	6
8	16
2	4

By that time, everybody thought they knew the rule, so they guessed that it was "times two." Somebody else said the numbers were doubled, but after talking it over they decided it meant just about the same thing, in different words.

The children and Aunt Bebe checked all the numbers, and the "times two" rules worked each time!

Aunt Bebe said another way to check the rule was to do the reverse, and see whether the answer came out **zero** if it involved addition or subtraction, or **one** if it involved multiplication or division. So they checked all the numbers this way, and that worked too.

☐ Tell your family that you have a calculator that is a lot like Aunt Bebe's, and that a rule has been lost inside it. If they will take turns giving you numbers to enter, maybe you can find out what the rule is.

☐ Each rule may include only one operation.

☐ As each person gives a guess, enter the number, then $\boxed{=}$, and make a record of the guess and the resulting number.

☐ When one of the players thinks he or she knows the rule, have them give you a number to enter and tell you what the result will be. Then try the number. Continue to let others guess until each person has had ten guesses or has found out the rule. Then give the rule.

☐ If some of the group is having trouble recognizing the rules, use a rule that is as simple as possible—something like +2, or ×2.

☐ When most people understand the idea, show everybody how to hide the rules in the calculator, and let them take turns being the hider.

☐ Talk together about the patterns, and how people guessed some of the hard rules.

The Lost Number

Why

To practice pattern recognition and to develop understanding of inverse relationships

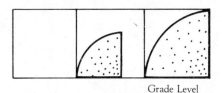

Grade Level

TOOLS

Calculator with a constant feature

▶ *An important part of comprehending the true nature of mathematics is being able to observe and interpret number relationships. One of the most important is the* **inverse** *relationship. For every number, there is another number which, when multiplied times the first number, will give the answer of 1. This is called the* **multiplicative inverse** *and is also called the* **reciprocal.**

For example, for 9, the number is 1/9, since $9 \times 1/9 = 1$.
1/9 is the multiplicative inverse or reciprocal of 9.

There is also another number which, when added to the first number, will give the answer of 0. This is called the **additive inverse.** *For example, for 9, the number is −9 or a negative 9, since $9 + (−9) = 0$.*

This activity uses the concept of a multiplicative inverse, with an intermediate step, so that the equation would read: $9 \times 1/9 = 9 \div 9 = 1$. A number is entered into the calculator so that it will divide other numbers, and when it divides itself, of course, the answer is 1. ◀

How

☐ Before you introduce this activity to your family, set up your calculator to use the constant divide feature. Practice with several numbers to be sure you know how to use the feature. For most machines, the constant divide can be set up by pressing the ÷ key, then a number, then = . For example, ÷ 7 8 2 = will set most calculators to divide by 782 any number entered.

☐ When you feel secure, "hide" a number in the calculator and read this story to your family:

> One day Aunt Bebe called all of the neighbor children together and told them that she had just discovered that there was a number lost in her calculator, but she needed their help in finding out what the number might be. She said the instructions that had come with the calculator said if a number got lost, the following steps should be taken:
>
> ☐ Enter a 3-digit number.
> ☐ Enter = .
> ☐ Write down the answer—this is a clue.
> ☐ Continue to enter 3-digit numbers and = to collect clues.
> ☐ The 3-digit number that gives an answer of 1 is **the lost number!**
>
> Well, it took Aunt Bebe and the children ten guesses to find the first lost number. How many guesses do you think it will take us to find the number lost inside this calculator?

☐ Now ask for guesses, and write them down so that everybody can see the list of guesses and resulting clues. Encourage working together to decide on guesses. Here is a sample game, with the hidden number 782:

Guess	Clue (shown on the calculator)
123	0.157289
100	0.1278772
999	1.2774936
888	1.1355498
777	0.9936061
788	1.0076726
780	0.997424
785	1.0038363
783	1.0012787
782	1.

☐ After the first game, talk together about how you used the clues to help find the hidden number. After a few games, show everybody how to hide numbers and have them take turns hiding and finding numbers. Encourage as much discussion as possible about ways to guess the number with fewest possible clues.

Pseudo-Monopoly

Why

To provide practice calculating percents and using calculators

▶ *This activity combines classroom mathematics with real-life applications, such as variable taxes, purchases, keeping neat records, and bonuses.* ◀

How

☐ **Read all of the directions before you start to play.**

☐ The game can go on for a long time, so you may want to set a time limit before beginning.

The Game

☐ Take turns rolling the two dice or spinning the spinner twice.

☐ Use the two numbers you roll or spin to form a two-digit number that represents the amount of money you have won on this turn.

For example, if you rolled a 2 and a 4, you may choose to win $24 or $42.

☐ But you **must** pay some TAXES!!!

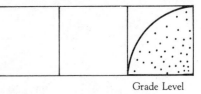

Grade Level

TOOLS

Paper and pencil

Pair of dice or a spinner with 1,2,3,4,5,6

Calculator

Cards with the numbers 2,3,5,7,11

Tax schedule

$11–$26 pay 15% of your number
$31–$46   25%
$51–$66   35%

▶ *To calculate percent, convert the percent to a decimal fraction (15% becomes .15) and multiply the decimal fraction times your number.* ◀

☐ Calculate your tax and subtract from your number.
  ☐ Round off after each operation.
  ☐ All dollar and cents amounts are rounded off to the nearest dollar.
  ☐ All 50¢ amounts are rounded up to the nearest dollar.

First roll $25 won   15% tax of $3.75 rounds to $4
                    −4 tax
                    ─────────
                    $21 won

□ For example, $21.32 becomes $21, $24.85 becomes $25, and $28.50 becomes $29.

□ Keep a neat recording of your accounts. See also the section on Record Keeping.

Primes

□ After the first round of play, you can buy a prime number whenever you have enough money. The prime numbers you can buy are 2, 3, 5, 7, or 11.

□ The prime numbers cost $500 times their **reciprocal**. For example, 7 costs $1/7 \times 500$, or $71.43, which rounds off to $71.

▶ A **reciprocal** *is the number that results when 1 is divided by a given number. The reciprocal of 5 is 1/5, of 36 is 1/36.* ◀

□ If anyone chooses to use a number **divisible** by a prime that is owned by another player, he or she must pay a commission of 50% of the winnings after taxes.

▶ **Divisible** *means that the number can be divided evenly by the prime, with no remainder.* ◀

□ If you need cash, you can sell your prime back, for its cost minus 10% interest.

Bonus

- ☐ If you roll a prime greater than 20, you get a 20% bonus before taxes.
- ☐ But you must pay taxes on the **new** number. For example, for the number 31, you would get a $6 bonus and you get $37. The tax would be 25% of $37, or $9.

Second roll	$31 won +6 bonus	20% prime bonus of $6.20 rounds to $6
	$37 −9 tax	25% tax of $9.25 rounds to $9
	$28	

Winner

- ☐ The winner is the player with the most money when the game is stopped, after all players sell back their primes.

Record Keeping

- ☐ All players must keep a neat recording of their accounts.
- ☐ Anyone may check another player's accounts before the turn passes and charge a $10 fee for correcting any errors found.
- ☐ Record in the following manner:

First roll	$25 won −4 tax	15% tax of $3.75 rounds to $4
	$21 won	
Second roll	$31 won +6 bonus	20% prime bonus of $6.20 rounds to $6
	$37 −9 tax	25% tax of $9.25 rounds to $9
	$28	
Total so far:	**$49**	$21 +$28 or $49

More Ideas

☐ Learn to use the following shortcuts and/or the percent key on your calculator.

☐ For a 20% bonus, $35 + (35 \times .2)$ is the same as 1.2×35

For a 15% tax, $35 - (.15 \times 35)$ is the same as $.85 \times 35$

What is the shortcut for a 25% tax? 35% tax?

☐ You can also use your percent key.

For example, for 20% of 35, press

$$\boxed{3}\ \boxed{5}\ \boxed{\times}\ \boxed{2}\ \boxed{0}\ \boxed{\%}$$ (without an $\boxed{=}$)

for 35 + 20% of 35, press

$$\boxed{3}\ \boxed{5}\ \boxed{+}\ \boxed{2}\ \boxed{0}\ \boxed{\%}$$;

and for 35 − 20% of 35, press

$$\boxed{3}\ \boxed{5}\ \boxed{-}\ \boxed{2}\ \boxed{0}\ \boxed{\%}$$

A Word About Microcomputers

Not since television has a gadget created as much excitement and worry as the personal computer. It is now possible to put onto your kitchen table the computing power that filled several rooms a few decades ago. Rumor has it that, if cars had evolved at the same pace as computers, a Mercedes would cost only fifteen dollars, have enough horse power to drive around the world in four hours, do it on one tank of gas, and be about the size of a pin head.

This mythical Mercedes would present some of the same problems we face with the microcomputer. Do we really want one? Sure, it's powerful, but what on earth would we do with it? What are its limitations? Are there hazards connected with its use?

And most of all, what do computers have to do with my children's lives?

At the time this book is being written, many aspects of personal computer usage are unclear. Different people are beginning to use computers for different tasks every day, and it seems that there is an explosion of new ideas of what can be done with these handy little machines.

Although there is a lot we don't yet know, there are a few things we can say with certainty:

☐ There will be computers used in schools, and it will be up to local communities (that's you!) to decide whether they should be used for drill and practice or for more creative learning, such as problem solving and programming.

☐ Almost every working adult within the next twenty years will probably have used a computer for some job-related task. Not everybody will be writing new programs, but charge accounts, bank accounts, investments, industrial designs, even perhaps school grading will involve computers.

☐ Computers will become less and less expensive and easier to use.

☐ Children should be given the chance to become familiar and comfortable with computers. This applies to **all** children, including girls and boys, children of all races, and all socio-economic groups.

What can you do to help your children?

First of all, become an expert yourself, or at least learn as much as you can. You will want to find out about the machines and about the commercial programs available. A good place to start is by reading. We strongly recommend the book **Parents, Kids and Computers** by Lynne Alper and Meg Holmberg, listed in the Resource List. If that is not available go to the public library, the school library, a computer store, or a book store, and see what other books are there.

You can also take courses at a local college or science center, or send your children to computer courses and learn from them. You may also ask your school to set up an evening when the computer lab or a classroom with a computer is open for parents to come and be taught by their children or by the teacher.

Since computers don't really do much by themselves, you will want to review a lot of educational software (the programs that make the computers do specific things). Some of the things you should look for:

☐ Is this activity worth having your child do?

☐ Could this task be done just as well or better using cheaper and more concrete objects like blocks, beans, pencil and paper?

☐ Are the values presented in this activity consistent with your own? Do you mind, for example, that your child's reward for solving a series of problems correctly is to play a game entitled SQUASH THE PIG?

☐ What is the quality of the interaction of the program with your child? Is she or he treated with dignity and respect? Is he or she learning to master the machine, or is the computer in control?

☐ Is the activity interesting?

☐ Are the directions and the supporting documents clear?

When you have a computer available, look back at the suggestions for helping your child with mathematics, earlier in this chapter. Children using computers also need to use language, work in groups, feel their parents are interested, and so on. Instead of trying to set up a highly organized program, you and your child should do a lot of exploring.

If you are teaching a FAMILY MATH class, perhaps you or a member of your class or some other friendly local expert will guide your group through initial computer encounters. These introductory hours might include software demonstrations, discussion about the various kinds of computers available and their merits and costs, explanations of the current major uses of computers in the business world as well as in the schools, a beginner's lesson in LOGO or some other computer language, or a time when you can browse through software on your own.

For the sake of yourself and your family, buy, borrow, or rent a computer and play with it. Go to computer stores, ask for information, and try out computers and software. Try writing a letter on a word processor. Try making a birthday card with one of the drawing programs. Explore this wonderful new tool. You may open a whole new world for yourself, and you surely will give your children an extra boost for the future.

Quick Questions Answers

(See page 229 for Quick Question instructions)

Answers

1) $8.00
2) 24 cases
3) 86,400 seconds
4) 2-1/4 years
5) 15 times per day
6) 36,000 times per day
7) 1800 messages
8) 92 hot dogs
9) 2% of microcomputer purchases
10) 359-12 oz. servings
11) 35 miles
12) 226,263

*Most of the answers to these Quick Questions are from the book, *American Averages* by Mike Feinsilver and William B. Mead, Dolphin Books, New York 1980.

Chapter XI

CAREERS

CAREERS

Information about careers is important for both parents and their children. Knowing the benefits and requirements for a great variety of future jobs helps plan the courses needed in school for a student to keep her or his options open. Students, especially female and minority teen-agers, often drop out of math when it becomes optional in high school. They think they will never have to use it. The suggestions and activities in this chapter will make students aware that math is used in many more jobs than they might expect.

Until recently, many women did not expect to work full time; they planned to work for a few years before marrying and perhaps again after their children had entered school. Now, the average married woman works 28 years of her adult life, while an unmarried woman will work 35 to 50 years on average—just as men do. All of our daughters, as well as our sons, must be prepared to enter the workforce and to have as many choices as possible. The activity "Useful Math Skills" answers the question "When are we ever gonna have to use this?"

"Where Are the Jobs?" will give students and their parents a realistic look at where the jobs really are, and whether they can reasonably expect to be movie stars or professional athletes. This does not mean that your children should not be able to choose those occupations too, but they should be prepared with alternatives if their choices don't work out.

Besides using the activities in this chapter, do try to have your children meet as many people as possible from as many occupations as they can. You may want to help the classroom teacher organize role model visits to the class, or arrange for field trips to nearby businesses. Ask the teacher whether parents from the class could come in and share something about their jobs, then contact parents and set up the schedule.

Information about careers can be a very powerful motivation to students to take classes they would otherwise drop.

A Working Day

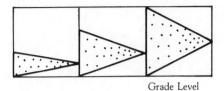

Why

To help parents and children think and talk about what it means to be a working adult

How

Paper and pencil

☐ Ask your children each to write a story (or essay if they are older) describing the kind of working day they think they will have when they grow up, beginning first thing in the morning and ending when they go to bed. Illustrations are nice, also. Have younger children dictate their story to you.

☐ When the stories have been written, read them carefully, and think about what you have learned about your children's aspirations. Does your child expect to have a job? a house? family responsibilities? Who washes the dishes in the imaginary family? Who drives the car? Do the children go to a baby-sitter, or does the mother of the family stay home all day? If you have several children, you may notice some differences in what they expect.

☐ Talk with your children about whether they think that is what really will happen, or whether it might be different from what they wrote. Some children will describe lives that are a lot like those of their parents, while other children will describe situations that are completely different.

☐ The purpose of this exercise is not to find out what your children really will be when they grow up, but to open the conversation between you and the rest of your family regarding all of the wonderful opportunities that may exist.

Career Story Cards

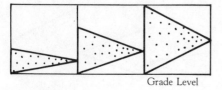

Grade Level

TOOLS

**Career story cards
(make a copy
pages 265-270)**

Scissors

Why

To initiate thinking about what different people do at work and to improve communications skills

How

☐ Have a copy made (at a duplicating center) of the Career Story Cards, on card stock if possible.

☐ Have your child cut out and sort the cards. There should be three stacks, showing **careers, where they take place, and what the person does.** This sorting can be a very interesting process, with a lot of talking about what the cards mean. The distinction between people, places, and actions is important in developing language skills.

☐ The cards may be spread out, so that the child can see them all.

Matching

☐ Have your child select a card that shows a person, then find a place card and an action card that might match, explaining why those cards make sense.

☐ There often will be more than one choice of cards that might match. For example, does a truck driver ever use a calculator? Probably—and so would almost any of the occupations shown on the cards.

Stories

☐ Have your child choose one card from each of the three categories and tell you a story about the person in that job, what the job was like on a particular day, and where it was done. Remember that all of the people also do other tasks than those they are pictured doing.

What's in a Job?

☐ Have your child take one person card, then see how many of the actions shown on the action cards might be done in that career, and how many of the places shown whould be likely.

More Ideas

Cut out pictures of people doing a variety of jobs, tools that are used, and places where work is done, from magazines. Make up a new set of career story cards using the pictures.

► *Here are some additional activities for* **classroom** *use:*

► *Have several children work together to make up the career stories.*

► *Have the children ask each other questions, using the cards to answer. For example, "Who uses numbers?" or "Who reads a graph?"*

► *Have the group put the cards into a long string, one at a time, explaining the connection between the last card and the new one put down. For example:*

 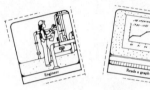

"Engineer"   "reads a graph   "and uses a   "and tells the
             of the ocean   calculator to   electrician where
             depth"   figure the   the wires
                               results"   should go"

► *Encourage students to make up their own activities with the cards.* ◄

Draws

Writes

Measures

Makes a report

Works outdoors

Uses a computer

CAREER STORY CARDS

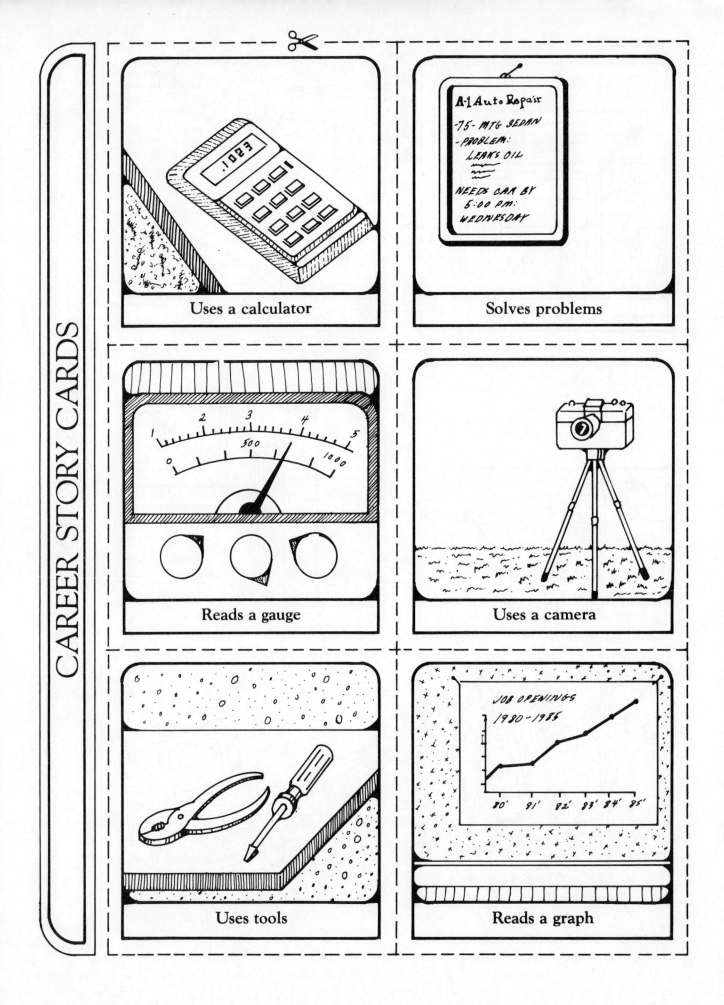

Uses a calculator

Solves problems

A-1 Auto Repair
-75- MTG SEDAN
-PROBLEM:
LEAKS OIL
NEEDS CAR BY
5:00 PM:
WEDNESDAY

Reads a gauge

Uses a camera

Uses tools

JOB OPENINGS
1980-1985

80' 81' 82' 83' 84' 85'

Reads a graph

Mechanic

Engineer

Truck Driver

Pharmacist

Computer scientist

Doctor

Construction project

Airport

Hospital

Shopping center

Skylab

Museum

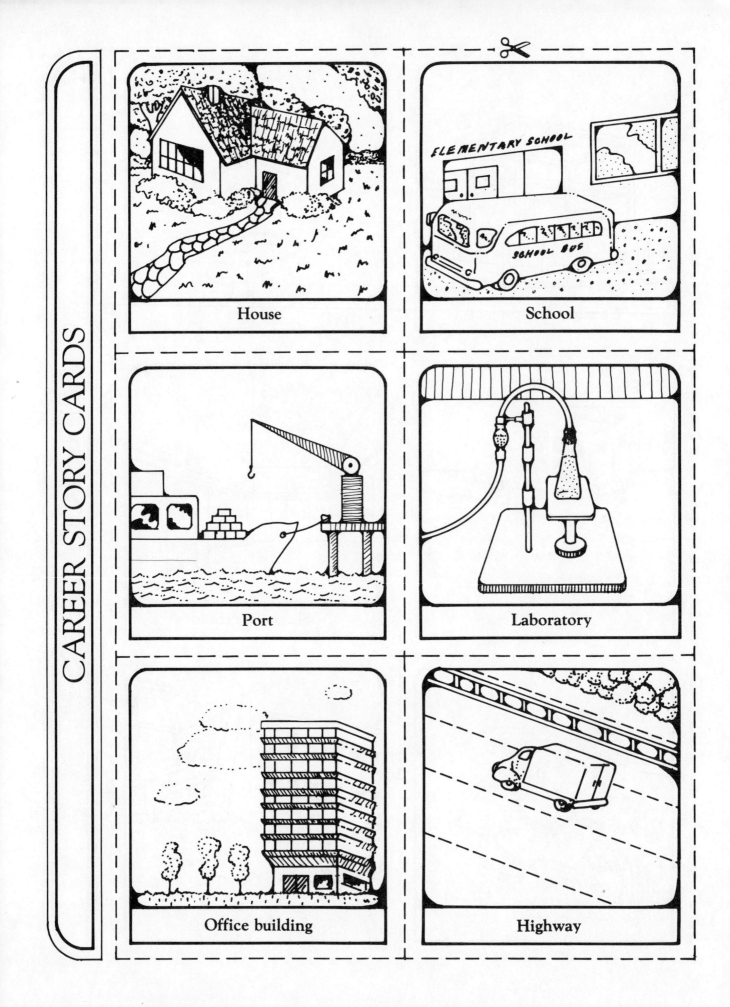

CAREER STORY CARDS

House

School

Port

Laboratory

Office building

Highway

Useful Math Skills

Why

To provide information about math as it is actually used on a variety of jobs.

How

☐ With your family, look at the following list of mathematics skills:

Fractions   Calculators

Basic geometric concepts   Formulas

Decimals   Averaging

Ratio and proportion   Estimation

Percent   Statistical graphs

Now, think carefully about all of the occupations that you know of. There is a list of 100 different occupations on the next page for you to look at if you wish. Imagine that you asked people from all of those occupations what math skill they use most often. Which of the above skills do you think it would be? Write that skill at the top of a piece of paper.

☐ Then decide which they might say is used next most frequently, and write it next on your list.

☐ Then decide which skill is used third most frequently.

☐ Continue until you have written all ten of the skills on your list. You may not all agree on the order. Talk it over and see whether you can come to some consensus.

☐ When you have come close to a general agreement, look at the answers on page 273. How close did you come?

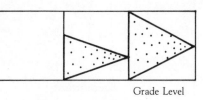

Grade Level

TOOLS

Paper and pencil

OCCUPATIONS

Accountant
Accounting Systems
  Analyst
Administrator:
  Shopping Mall
Advertising Agent
Airline Passenger
  Service Agent
Airplane Mechanic
Airplane Pilot
Air Traffic Controller
Appraiser (Land)
Architect
Artist (Graphic)
Attorney
Auditor
Auto Mechanic
Bank Teller
Biologist (Environmental)
Carpenter
Carpet Cleaner
Cartographer
Chiropractor
Computer Programmer
Computer Systems
  Engineer
Contractor (General)
Controller (Hospital)
Counter Clerk
  (Building Materials)
Data Processor
Dentist
Dietician
Doctor (G.P.)
Drafter
Economist
Electrician
Electrical Engineer
Electronics Technician

(Civil) Engineer
(Electronics) Engineer
(Industrial) Egineer
(Petroleum) Engineer
Environmental Analyst
Farm Advisor
Fire Prevention Officer
Fire Fighter
Forestry Land Manager
Forestry Recreation
  Manager
Geologist
  (Environmental)
Highway Patrol Officer
Hydrologist
Income Tax Preparer
Insurance Agent
Insurance Claims
  Supervisor
Interior Decorator
Investment Counselor
Landscape Architect
Librarian
Machinist
Manager:
  Appliance Store
Manager:
  Temp. Employment
  Service
Marketing Rep.
  (Computers)
Masonry Contractor
Medical Lab Technician
Meteorologist
Motorcycle Sales and
  Repair
Navigator
Newspaper: Circulation
Newspaper: Production
Newspaper: Reporter
Nurse

Oceanographer
  (Biological)
Optician
Orthopedic Surgeon
Painting Contractor
Payroll Supervisor
Personnel Administrator
Pharmacist
Photographer
Physical Therapist
Plumber
Police Officer
Political Campaign
  Manager
Printer
Psychologist
  (Experimental)
Publishing:
  Order Manager
Publishing:
  Production Manager
Purchasing Agent
Radio Technician
Real Estate Agent
Roofer
Savings Counselor
Sheet Metal/
  Heating Specialist
Social Worker
Stock Broker
Surveyor
Technical Researcher
Title Insurance Officer
Travel Agent
T.V. Repair Technician
Urban Planner
Veterinarian
Waitress/Waiter
Wastewater Treatment:
  Operator

► *This list is based on a survey of one hundred people from the 100 different jobs shown below. Each person was asked what kind of mathematics he or she actually used on the job. The person who made the survey was Hal Saunders, a junior high school math teacher in Santa Barbara, California. The information was first published in an article, "When Are We Ever Gonna Have To Use This?" written by Hal Saunders for* **The Mathematics Teacher,** *January, 1980. He points out, however, that since only one representative of each occupation was interviewed, more data may be needed.* ◄

More Ideas

Have the members of your family make their own survey of people they know, asking them what math skills they use on their jobs.

Useful Math Skills—Answers

Math skill	% of the 100 jobs that use this skill	Rank
decimals	100	1
calculators	98	2
percent	97	3
estimation	89	4
fractions	88	5
averaging	83	6
ratio and proportion	77	7
statistical graphs	74	8
formulas	68	9
basic geometric concepts	63	10

Where Are the Jobs?

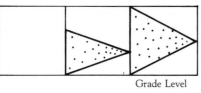

TOOLS

Paper and pencil

Why

To give families information about the occupational categories that reflect the composition of the workforce in the United States

How

☐ Write the following categories on a piece of paper:

1982 U.S. Workforce	%
Business	_____
Professional/Technical	_____
Health/Medical	_____
Industry/Transportation/Farms	_____
Service	_____
Communication/Art/Design/Glamour	_____

☐ With your children, estimate what percent of the U.S. Workforce works in each category. Imagine that there are 100 people standing in front of you. How many of the 100 would work in each category?

☐ When you have completed your list, look at the answers on the next page.

More Ideas

Since these figures were published in 1982, there have been other projections of increases and decreases in certain job categories. At the public library, you and your family may be able to look up what those changes were. Ask the librarian to help you find several places to look for the information.

WHERE ARE THE JOBS?

Answers

1982 U.S. Workforce

Business   35%
  Managers (11%)
  Sales workers (6%)
  Clerks/Secretaries (18%)

Professional/Technical   11%
  Scientists/Technicians (6%)
  Teachers/Social Workers/
    Librarians (5%)

Health/Medical   7%
  Doctors/Dentists/
    Veterinarians (3%)
  Nurses/Medical Assistants (4%)

Industry/Transportation/Farms   33%
  Skilled craft workers (12%)
  Operatives/Truck drivers/
    Machine operators (14%)
  Farmers/Laborers (7%)

Service   13%
  Police Officers/Firefighters/
    Waitresses/Beauticians

**Communication/Art/
Design/Glamour**   1%
  Entertainers/Models/
    Professional athletes/
    Fashion designers

These figures were taken from 1982 statistics provided by the U.S.
Department of Labor, Bureau of Labor Statistics, Washington, D.C.

Notes From An Engineering Aide
Who Teaches FAMILY MATH

This essay is included for parents and older children to read. It points out connections between FAMILY MATH activities and engineering work. In a way, it is another answer to "When are we ever gonna have to use this?" and explains why FAMILY MATH dwells more on problem-solving activities rather than memorizing arithmetic algorithms:

I work for Chevron Research Company as an engineering aide. Recently I was given a problem to solve. I'd like to talk about how I used the ideas presented in this class to solve the problem.

In the Richmond refinery there is a plant where used oil barrels, the big 50 gallon type, are reconditioned and reused as new barrels. To do this, the plant strips off the old paint using a strong detergent, washes the barrels and then paints them. The waste water from the plant is really dirty. It's filled with detergent, paint and oil. My job is to find a way to clean the waste water.

How does all this fit in with FAMILY MATH? Remember the first class when you were asked to estimate? There was the estimation sheet and the jar filled with beans. Well, the first thing I had to do was to make an estimate of how much waste water we needed to clean every day. I took a big bucket and went out to the plant. I put the bucket under one of the waste water pipes that emptied into a big basin in the ground. I took a stop watch and measured how long it took the water from the pipe to fill the bucket. In this way I was able to estimate how much waste water the plant produces every day. This is the same as counting a handful of beans to help you figure out how many beans there are in the whole jar.

Another thing I had to do was measure. I already mentioned that I had to measure the time it took to fill my bucket. I also had to take little samples of the waste water to the laboratory and measure how much of a certain chemical was needed to clean the water. I used milliliters and liters to do my chemical measurements. Your Mom probably uses teaspoons and cups to measure what she needs when she makes cookies. You used a piece of string to measure how many times your height will go around your wrist, waist, and head. Remember the evening we spent measuring things with paper clips, safety pins and other objects. It doesn't matter what you do your measuring with, as long as you can keep track of what you have measured.

How about the trading game? We traded 4 blue strips for a purple one, four purple ones for an orange one. I had to do that too. When I did my work in the laboratory, I measured the chemicals in milliliters. The chemical that I needed is sold in pounds. I had to trade milliliters for liters and liters for pounds and pounds for dollars to figure out how much the chemical needed to clean the water would cost.

Tonight we worked with probability. I have to work with probability too. I had to take only three samples of waste water and do tests on them and use the results to make a decision about how to treat all of the water. The water changes every day. Sometimes it is so dirty it's like a thick milkshake. Other times it is mostly blue from the paint but isn't very dirty otherwise. I did many tests on the three samples. I used probability to decide what the best way to clean all the water might be.

Another idea I want to talk about is strategy. The balloon ride is a game that can be won every time if you know how it works. Tic-tac-toe is the same way. When I started working on my problem on the job, I had no strategy. I had the answer. I needed to clean the water. But how? Part of my job was to find a strategy that would help me solve the problem. I did this by talking to people who were helping me and organizing my time so I would be able to do all the estimating, measuring, probability studies and other chores before I came up with an answer. When mom bakes cookies, she probably uses a cookbook as a strategy to solve her problem of how to make cookies. It's important to have some kind of strategy before you try to solve a problem. It might not always be the best strategy, but it's better than none at all.

Tangrams are fun. They're neat too because they show you that there are different ways of putting things together. I had to figure out the best way to put a water cleaning system together that would work and didn't cost much. When you put all the tangram pieces of the tangram together just right, you make a perfect square. It takes a long time to do that and a lot of fiddling around to get it right. In the same way, I had to fiddle around with all the pieces of information that I had collected from my work and come up with a plan that would work better than any other. It was hard and I needed lots of help from other people. We talked a long time before I came up with an idea that we think will work.

These are some of the ways I used the same ideas that you use in FAMILY MATH at my job. If you pay attention, maybe you can see how you use them in your work too. It's not hard to do. You

just have to pay attention to what you are doing. Your mom and dad or some of your friends can help you with some of these techniques just as you can help them.

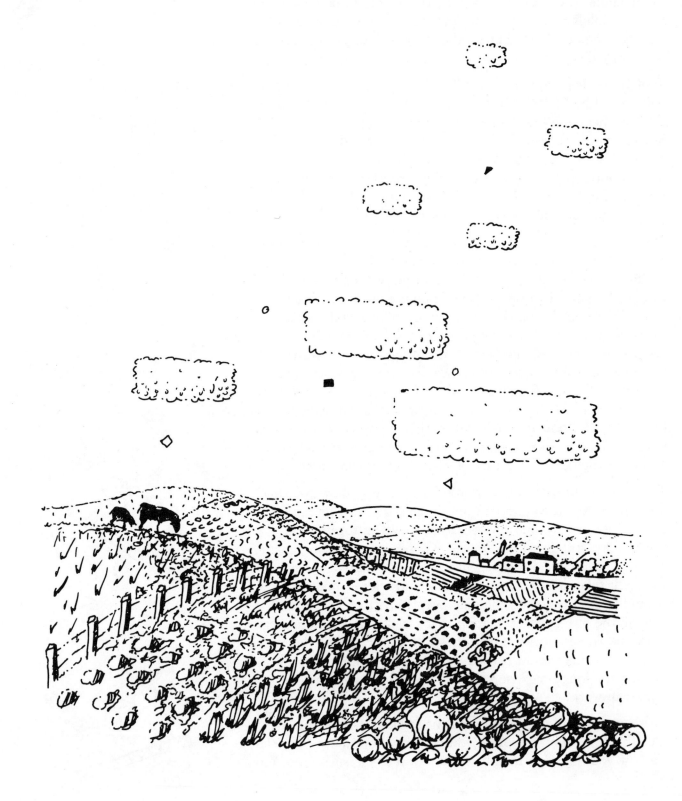

Organizing a Family Math Class

ORGANIZING A FAMILY MATH CLASS

Good news! There's no one way to organize and teach a FAMILY MATH class. The class can be taught by parents, teachers, administrators—the only requirement is a desire to share mathematics in an enthusiastic and nonthreatening way. We've included a number of suggestions gained from our own and others' experiences in starting the program, but you are the expert on what will work best in your own community.

WHO SHOULD COME?

Grade levels An early decision involves what grade level(s) you want to teach. Some people prefer offering a class for only one or two grade levels, since they think it's easier to present activities that will hold the interest of all the children. On the other hand, this is **family** math, and parents might want to bring an older or younger sibling. Many classes have been successful addressing a K-6 or 4-8 range. Pick what suits you best, based on your interest and knowledge of the math topics and the needs of your community.

WHEN IS THE BEST TIME?

Scheduling There seems to be greater interest in attending FAMILY MATH classes during the fall or winter months than in spring. Naturally, you'll want to avoid school vacations, holidays, and other busy times. As to time of day, we have found it very difficult to maintain an afternoon class, but the early evening has nearly always been successful. Sometimes a Saturday morning class has worked. Before you set a time for your class, you may want to send home a recruitment notice with the students you are targeting, including a return portion for parents to indicate their preferred time.

Here is a sample form:

FAMILY MATH

Learn

* ★ How to help your child with math at home
* ★ What math your child will learn in school this year
* ★ How to make math fun

Take Home

* ★ Materials; games and activities; career information

Meet

* ★ Women and men who use math in their jobs

For more information, call:

Please return the bottom of this sheet to:

- -

☐ **Yes, I'm interested in FAMILY MATH.**

Name

Address

Telephone

Best time to call

Child's name

Child's school

Child's grade

Child's teacher

Best time for me to attend a FAMILY MATH class:

Weekdays: (1:00– 3:00)

(3:00– 5:00)

(6:00– 8:00)

Saturdays: (10:00–12:00)
First choice

Second choice

☐ I cannot attend a FAMILY MATH class, but please keep me on your mailing list.

WHERE IS THE BEST PLACE?

Location   A school classroom or library, church, community center, YMCA, YWCA, or science center are all good places for a FAMILY MATH class. Some things to look for, if possible:

☐ No charge or only a small fee for the use of facilities

☐ Enough tables and chairs for everyone, with some for extra activities

☐ A separate room for very young children, with a babysitter if possible

☐ Ample parking

☐ Safety for night classes

HOW DO I GET PEOPLE TO COME?

Recruiting   This can be the easiest or the hardest part of the class. If you know what a typical turnout is at your school for open houses, PTA meetings, and school plays, you may have some idea how difficult or easy it will be. If your community typically has high parent involvement, you may have more people interested than you can handle. If the parents in your community do not attend school functions but are regular church-goers, your best strategy may be to recruit through churches, or you may include parents from other classes or schools.

Notices   If you want to work with families of K-6 children, a good procedure is to ask the teachers of those classes to send home notices with their students, similar to the sample. Parents of middle and junior high school students will need more encouragement than parents of younger children, and notices may need to be mailed. And don't forget about the telephone. You may want to emphasize in your notices that FAMILY MATH will provide parents with a chance to learn about their children's math curriculum, practice problem-solving skills with their children that will help them succeed with this curriculum, and encourage them to persist later as math becomes optional.

Parents and children   It's important to be very clear that your FAMILY MATH classes are for parents (or other interested adults) who may attend with or without a child, but that NO CHILD MAY ATTEND WITHOUT AN ADULT. If you do not insist on this rule, you may find yourself with a very successful tutorial session for a lot of students who want extra help or enrichment activities. This is not the purpose of FAMILY MATH.

Be aware that it is possible to over-recruit. Decide how many people you are willing to work with; and if you think you might be overwhelmed with responses, start small, recruiting from parents of one or two classes.

Don't forget radio and newspaper announcements if you want to reach a large audience. Be sure to include a telephone number.

Sample Radio Spots

9 February 84

FOR IMMEDIATE RELEASE (class starting February 23)

Public Service Announcement—30 seconds

IF YOU'RE THE PARENT OF A JUNIOR-HIGH-AGE SON OR DAUGHTER, YOU SHOULD KNOW THAT ONE OF THE MOST IMPORTANT ACTIONS YOU CAN TAKE RIGHT NOW, TO HELP THEM GET GOOD JOBS WHEN THEY GET OUT OF SCHOOL, IS TO MAKE SURE THEY KEEP ON TAKING MATH WHILE THEY'RE **IN** SCHOOL. EVEN IF YOU FEEL LIKE A BONEHEAD YOURSELF IN MATH, YOU CAN HELP YOUR TEENAGER. COME TO A FREE FAMILY MATH CLASS WEDNESDAY NIGHTS, STARTING FEBRUARY 23, AT CALVIN SIMMONS JUNIOR HIGH SCHOOL, 35TH AVENUE AND FOOTHILL IN OAKLAND. FOR INFORMATION, CALL 642-1823.

9 February 83

FOR IMMEDIATE RELEASE (class starting February 23)

Public Service Announcement—10 seconds

OAKLAND PARENTS OF JUNIOR HIGH STUDENTS! HELP YOUR KID SUCCEED IN MATH! COME TO FREE "FAMILY MATH," WEDNESDAY NIGHTS AT CALVIN SIMMONS JUNIOR HIGH. FOR INFORMATION, CALL 642-1823.

Group presentations

Certainly, one of the most powerful ways to attract families to your class is to offer to make a presentation at a PTA or other parent group on on "How to help your child with math at home." In twenty minutes you can explain the program, share suggestions for math activities that are easy to do at home, and do at least one math activity with the group. Collect names and phone numbers of interested parents. *Value of Words* (see page 33) is a good activity to do in a short presentation because it is interesting to adults as well as children and gets everyone involved. Be sure to bring copies of the activity you present, enough for everybody, with flyers to inform them about the class. Another good use of a longer time is to show the FAMILY MATH film, "We All Count." This seventeen-minute film or video gives a good view of what different FAMILY MATH classes look like and why they're taught. The film is available from the Lawrence Hall of Science. Write to the address inside the front cover for further information.

FINANCIAL SUPPORT

Finances Make a list of your possible expenses to be sure you can cover your costs. Although it is possible to charge a "per family" fee, it should be kept as small as possible. Don't be afraid to ask for donations of all kinds—principals, businesses, philanthropic organizations, parent groups, your local grocery store or gas station—most people find the idea of FAMILY MATH very exciting, and will be willing to contribute materials or money.

Expenses Here are some of the expense items to consider:

☐ Handouts and materials—probably about $5.00 per family for a six-week session.

☐ Rent—try to get space free in a school or church.

☐ Refreshments—you may be able to get them donated, or parents may be willing to take turns bringing them.

☐ Your own time—much of your time will be donated, but you may be able to arrange a small salary through adult school or community college sources.

SETTING UP

Preparation See the Planning Check Sheet on page 291. Be sure you have all the details arranged, including enough handouts for everybody, all the materials for each activity, and a seating arrangement that is comfortable for you and the participants to see and hear each other.

Set-up Give yourself plenty of time to set up—be ready for the early birds. For the first session, you may need to arrive an hour or more before the official time. To help everyone get to know each other, provide name tags. Have juice or coffee available. If you want to keep attendance—and it's a good idea for record keeping—ask people to sign in on a Venn Diagram:

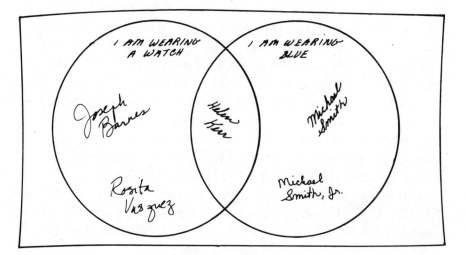

You can explain that a Venn Diagram is a way of sorting and classifying information and that these diagrams can be very simple or complex. (A series of Venn Diagram activities can be found on page 59.) You may also want to maintain a more formal attendance list, with addresses and phone numbers so you can keep in future contact with your families.

OK, THEY'RE ALL HERE—NOW WHAT DO I DO?

Be warm, be nurturing, be energetic...the atmosphere that you establish is the most important part of the class. A nonthreatening, comfortable environment gives parents and children the confidence to take some risks, try something new, make mistakes. The activities themselves will encourage discussion, and the way you encourage question-asking and talking will help people feel free to speak out.

The most important messages you want to convey to parents are not necessarily put into words, but are conveyed through your own actions—that mathematics has great beauty, that it can be learned by everyone, and that it can be a pleasure to do mathematics. When people can begin to concentrate on the fun of doing mathematics rather than getting the right answer, your class will be a success.

Welcoming

LET'S BEGIN

A typical FAMILY MATH class begins with an activity that can be set out with little or no explanation as an "opener" for people to do as they arrive. It's important to begin the class at the designated starting time, but you don't want parents to feel uncomfortable if they are late. You can begin a discussion of the opener about five or ten minutes after the official start of class. Some activities that work particularly well as openers have been gathered in a separate chapter called "Beginning."

Class openers

At your first meeting, you'll want to explain the FAMILY MATH program, make introductions, and give an overview of your series of classes. It is probably best, for your first activity, to pick something that is based on arithmetic skills, since this will be familiar and acceptable. Later, when you introduce an activity from a less familiar math topic, you can talk about the interrelationships of the math topics you are covering.

Program description

After each activity, take time for questions and discussion. Save your most challenging activities for a later class because you want everyone to be successful in the first activities. Many adults are very intimidated by math, and just coming to this class can be a big step for them. You want to make sure they'll come back next week!

Discussion

Stations

Sometimes "learning stations" can be used as a collection of openers. These are short activities arranged around the room with self-explanatory instructions, giving families an opportunity to apply problem-solving skills and strategies in a more independent situation. Measurement activities work especially well for this. Stations allow people to spend more or less time on any activity, setting their own pace.

While families are working at the openers or stations, circulate among them providing encouragement and suggestions where needed. Make observations of points to be discussed later. When you close this period of the class, explain the mathematics required, why these activities were chosen, and how they fit into the curriculum.

Lesson plans

Sample lesson plans are provided on pages 292 to 296 to help you pick and choose how to organize your class. At the end of the class, you might ask people to write their comments about the evening—if you provide them with paper and tell them they don't need to write their names on the comment sheets, you'll probably get a very candid group of comments. This will help you revise and refine your next class.

Handouts

Be sure to give the families copies of all the activities they've done in class and reminders about using them at home during the week. At your next class meeting, discuss the activities the families did at home. Encourage parents (and children!) to keep journals of their experiences as they explore mathematical ideas.

THE ACTIVITIES

Topics

The activities in this book are arranged by mathematical topic. In selecting the activities you want to use, you may decide to teach from one topic per class, or a selection of topics each time. Since there are not usually clearcut lines separating the activities, some will belong in several places. All the activities in the book have been chosen because they promote problem solving and mathematical reasoning, they involve the use of concrete materials, and they can be enjoyed again and again. Most of the activities can be interesting to a large age span and parents are often adept at adjusting the activity for both older and younger children.

Directions

Spend some time at the beginning introducing each new activity with clear directions. Point out the mathematical skills involved and how the activity relates to the school curriculum. You may want to stop the activity about halfway through to talk about strategies they are using, patterns they see, or predictions they want to make. Your goal is to guide them to their own solutions and strategies, stressing the importance of the process rather than simply getting the right answer.

> *This is one of the greatest gifts you can give parents and children—the understanding that getting **to** the answer is not as important as understanding **how** to get there. This message should be repeated often throughout the course.* ◄

As you close the activity, repeat its connection to the curriculum and the math skills involved. You may want the families to brainstorm the skills and strategies they have used to solve the problem; you may want to bring closure to the problem by having the whole group work all the way through a solution process; or you may decide to let the families work it through on their own.

No matter how you organize or teach your class, it is important always to provide written activity instructions for the families to refer to at home. You also may want to include copies of gameboards, special dice, etc., in the materials you send home, although we have tried to select FAMILY MATH activities that use inexpensive, everyday materials that can be found at home.

IN OTHER LANGUAGES

FAMILY MATH can be an excellent vehicle to reach families who do not speak English. It provides a comfortable way for them to become involved with school and to learn the importance of mathematics in this culture. If the instructor does not speak the second language, a translator will be needed, preferably one who is interested in or understands the activities. Sometimes other bilingual parents may be able to provide translations. Be sure to allow time for these conversational translations to be completed before moving on too quickly.

It is important to provide translations of the handouts if possible, and to make sure all questions are answered. Check with the translator frequently. There may be many questions, since this way of approaching mathematics may differ drastically from the traditional teaching of many countries. Your explanations will be important in encouraging the families to return each week.

SPECIAL NOTES FOR A JUNIOR HIGH SCHOOL OR MIDDLE SCHOOL CLASS

The school curriculum in the middle years leaves the familiar world of arithmetic and plunges children into the structure of arithmetic, geometry, and algebra. It is no longer enough to know that $1/4 \div 1/2 = 1/4 \times 2/1$; it now becomes important to know how division is linked to multiplication. Questions like 26×5 are replaced by directions like "Find all the values of the digit **d** such that 2 is a factor of 41**d**." There is often a tremendous push to

make everything symbolic and abstract, and parents, along with their children, become less sure of how to cope. In FAMILY MATH classes, parents and their kids can learn with less pressure.

Recruiting Recruiting may be difficult for these classes. Even though parents would like to help their children, they tend not to "interfere" with the school. You will need to make parents comfortable and to convince them that coming to a FAMILY MATH class will give them greater understanding of the curriculum and of the mathematics their children are learning, and that it will really help their children succeed. Work with teachers and counselors to try to get this message across.

Special topics At some point, you may wish to ask parents whether there is anything in the middle school curriculum that they never really understood but wished they had. You can then devote part of a session or two to exploring some of the frequently named messy math mysteries, such as figuring discounts, or compound interest, that are likely to be requested.

Careers Be sure in a middle school class to include career activities. These will give real strength to your suggestions that students should keep taking mathematics. Role models, or people from various occupations who visit your class and talk about their jobs, are a must.

Languages Send home recruitment notices in the language of the parents you want to attract. Word-of-mouth and sample activities may also help. In any case, you will find this kind of class among the most satisfying to you personally, as you watch parents and children enjoy mathematics together.

INTRODUCING CAREERS

Careers An important part of FAMILY MATH classes is information about careers, including role model speakers who visit the class. When we developed our first FAMILY MATH class, we wanted to have something for adolescent family members who were not regularly attending the class. Knowing that junior high and high school students are beginning to wonder (and worry) about The Future, we wanted to have them meet with young men and women working in a range of fields (from the skilled trades to business to research scientists, for example) to demystify some of the questions about their future choices. We made sure that our role models understood our emphasis on the importance of mathematics to future options, so they were able to tell personal stories about their experiences learning and doing mathematics.

Role models A panel of three or four role models is often the most effective way to plan this portion of the class. High school or college students majoring in math- or science-based fields are also excellent choices, particularly if they have attended school in the community.

To find role models, ask members of the class, your friends, or contact the local high school, especially the career center, and organizations at the local college such as the Society of Women Engineers.

Start the role model class with an opener, to give everyone time to arrive. When the panel begins, ask each role model to talk for about five minutes about the work they do or the field they are studying and how they decided to choose that field. Suggest that they talk about how parents and teachers influenced their choices. Hold all questions until all panelists have spoken. Be prepared with a math activity in the rare event that the questions run out before the class is over.

EVALUATION

It might seem strange to think about evaluation in such an informal setting, but there are a number of benefits in knowing what you've accomplished, for whom, and how you might improve the class. Here are our thoughts on evaluation....

Why should I evaluate?

- ☐ For my own information, to help me to improve the class
- ☐ For administrators, school board, community interest
- ☐ For potential supporters, who like to know what they're supporting

What could I evaluate?

- ☐ Who comes? Why did they come? What do they hope to accomplish? Who doesn't come, or doesn't come back? Why didn't they?
- ☐ How is the class going? Are people enjoying it? Are my presentations and explanations clear? Am I making the connections that people need?
- ☐ What are people taking home from the class? Are they more positive about math, about teaching math, and about their children's math education?
- ☐ What are people doing between classes? Do they use any of the activities? Do they find more math activities in their family lives? Are they getting more involved with their children's math education?
- ☐ What are some long-term outcomes of the class? Are parents more involved with the school? Are parents learning more math themselves? Are children more interested in math?

How can I evaluate?

- ☐ *Informal comments*—begin the class with introductions and a question, like: "How did you hear about the class?" "Why did you come?" "Did you try any activities from last week?"
- ☐ *Journals*—ask parents to keep notes of the classes in journals where they also record math activities and reflections they have between classes; borrow these journals to read at the end of the sessions.
- ☐ *Checklists*—list the activities you've presented by name and ask parents to rate them (didn't attend or don't remember, liked it, didn't like it; tried it at home, plan to try it sometime, don't plan to try).
- ☐ *Follow-up strategies*—talk to children's teachers, talk to children, have a follow-up meeting of parents.

Any other advice?

- ☐ Respect parents' and children's privacy, don't let others see their comments, gather only data that you plan to use, ask questions for a reason and try to act on people's recommendations if you agree that they're sound.
- ☐ Listening carefully is one of the best evaluation techniques available!

BEST OF LUCK!!

We hope that this section of advice on how to set up a FAMILY MATH class has been helpful to you. We have tried to give you enough information to proceed without overwhelming you with ideas. We'd love to hear how your class went. If you have questions or comments, please contact FAMILY MATH at the Lawrence Hall of Science, University of California, Berkeley, CA 94720, (415) 642-1823.

FAMILY MATH
PLANNING CHECK SHEET

Things to do before class:

When	What
1 or 2 months before class	☐ Decide on time and place ☐ Decide on grade levels ☐ Make arrangements with principal, district office, custodian and/or whoever else needs it
About 6 weeks before class	☐ Begin recruiting (Much earlier than 8 weeks and people may forget, and later than 3 weeks doesn't leave enough time for parents to plan)
1 or 2 weeks before each class	☐ Finalize class curriculum ☐ Begin to gather needed materials ☐ Prepare masters for handouts ☐ Line up child care if appropriate
About 2 weeks before the role model panel	☐ Select date and line up role models for career seminar
1 week before class	☐ Run off handouts for week 1 (guess on enrollment) ☐ Double-check room availability ☐ Send home reminder notices to those signed up
1 day to 2 hours before class	☐ Triple-check room arrangements ☐ Set up openers, sign in sheet. ☐ Arrange furniture the way you like it ☐ Make coffee, tea, etc. arrangements
When class begins	☐ Relax, its going to be wonderful!

Generic FAMILY MATH Lesson Plan

TIME	ACTIVITIES	HANDOUTS
10-15 MINUTES	☆ OPENER(S) _____ _____ _____	~ _____ ~ _____
5-20 MINUTES	~ INTRODUCTIONS ~ OVERVIEW OF EVENING (AND COURSE) ~ GO OVER OPENER ~ DISCUSS LAST WEEK'S HOME-WORK ACTIVITIES	~ _____ ~ _____ ~ _____ ☆
15-20 MINUTES	☆ MATH ACTIVITY _____	
5 MINUTES	DISCUSSION	MATERIALS
15-20 MINUTES	☆ MATH ACTIVITY _____	~ _____
5 MINUTES	DISCUSSION	~ _____
15-20 MINUTES	☆ MATH ACTIVITY _____	~ _____
5 MINUTES	DISCUSSION	~ _____
10 MINUTES	~ REVIEW ACTIVITIES TO BE DONE AT HOME AND SPECIAL HOME-WORK ASSIGN-MENTS ~ COMMENTS AND EVALUATION	~ _____ ~ _____ ☆

~NOTE: THIS SCHEDULE IS MEANT TO BE FLEXIBLE. YOU MAY WANT TO GO OVER HOMEWORK ACTIVITIES LATER IN THE CLASS. FOR SOME CLASSES YOU MAY WANT TO WAIT TO DISCUSS THE ACTIVITIES ALL AT ONE TIME. THE SAMPLE LESSONS THAT FOLLOW WILL ALSO ILLUSTRATE OTHER CIRCUMSTANCES

~

TIME	ACTIVITY	HANDOUTS
7:00–7:15	☆ GUESS AND GROUP	~ GUESS AND GROUP
7:15–7:30	~ INTRODUCTIONS	~ NUMBER ACTIVITIES FOR YOUNGER CHILDREN
	~ OVERVIEW OF LESSON AND CLASS	~ SORTING AND CLASSIFYING
	~ DISCUSSION OF GUESS AND GROUP	~ BALLOON RIDE
7:30–7:40	☆ ODD OR EVEN	~ K–1 CURRICULUM
7:40–7:45	~ DISCUSS COUNTING	☆
	☆ ON THE DOT	
7:45–8:00	☆ ANIMAL CROSSING	

		MATERIALS
8:00–8:15	☆ SORTING AND CLASSIFYING	
8:15–8:30	☆ BALLOON RIDE	
8:30–8:45	~ REVIEW ACTIVITIES FOR HOMEWORK	~ VARIOUS TYPES OF OBJECTS
	~ DISCUSS HOW TO WORK AT HOME	~ ODD/EVEN STRIPS
	~ GIVE OUT K–1 CURRICULUM SHEET	~ ON THE DOT PAPERS
	~ GIVE OUT INITIAL QUESTIONNAIRE	~ ANIMAL CROSSING BOARDS AND MARKERS
	~ COMMENTS AND EVALUATIONS	~ OBJECTS TO SORT
	☆	~ TOOTHPICKS
		☆

TIME	ACTIVITY	HANDOUT
7:00 - 7:30	~ MEASUREMENT ACTIVITIES AS STATIONS FROM LENGTH, AREA, VOLUME, CAPACITY AND WEIGHT	~ MEASUREMENT ACTIVITIES
		~ BLANK CALENDAR SHEETS
7:30 - 7:40	~ OVERVIEW OF MEASUREMENT FROM INTRODUCTION TO MEASUREMENT	~ PAYING THE PRICE
		☆
7:40 - 8:00	☆ MAKING A CALENDAR	
8:00 - 8:15	☆ PAYING THE PRICE	
8:15 - 8:30	~ REVIEW ACTIVITIES FOR HOMEWORK	
	~ DISCUSSION OF LAST WEEK's HOMEWORK ACTIVITIES	
	~ COMMENTS AND EVALUATION	**MATERIALS**
	☆	~ MEASUREMENT MATERIALS FOR ACTIVITIES
		~ CRAYONS OR MARKERS
		~ PENCILS
		~ REAL OR PLAY COINS FOR EACH PAIR OF PARTNERS 50 PENNIES 10 NICKELS 5 DIMES 2 QUARTERS ASK PARENTS TO BRING THESE TO CLASS
		~ SCRATCH PAPER

TIME	ACTIVITY	HANDOUTS
7:00-7:15	☆ TWO DIMENSIONAL NIM	~ TWO DIMENSIONAL NIM
7:15-7:35	☆ FRACTION KIT	~ FRACTION KIT
7:35-7:55	☆ FRACTION GAMES	~ FRACTION KIT GAMES
7:55-8:05	☆ COORDINATES I	~ COORDINATES I
8:05-8:20	☆ HURKLE	~ HURKLE
8:20-8:30	~ HAND OUT RESOURCE LISTS	~ RESOURCE LISTS
	~ DISCUSS WHERE TO FIND MORE ACTIVITIES AND CLASSES LIKE FAMILY MATH	~ EVALUATION FORMS
	~ FINAL EVALUATION	☆
	☆	

MATERIALS

~ GRAPH PAPER

~ CRAYONS

~ CONSTRUCTION PAPER FOR FRACTION KITS

~ FRACTION DICE

~ HURKLE PAPER

~ MARKERS

☆

Sample Lesson Plan

Middle School

TIME	ACTIVITY NAME AND BRIEF DESCRIPTION	HANDOUTS	OTHER MATERIALS
BEFORE CLASS	<u>TEN CARD ARRANGEMENT</u> ~ USED AS OPENER ~	NONE ~	CARDS NUMBERED FROM 1-10 (CUT INDEX CARDS TO MAKE SEVERAL PACKS) ~
	REFRESHMENTS	NONE ~	COFFEE ~ TEA ~ JUICE ~ PEANUTS ETC. ~
10-15 MINUTES	CLASS OVERVIEW ~ PRESENT THE SESSION SCHEDULE ON CHALKBOARD OR OVERHEAD. EXPLAIN THE RELATIONSHIP OF TEN CARD ARRANGEMENT TO MIDDLE SCHOOL MATH (LOGICAL THINKING ~ STRATEGY DEVELOPMENT, ESPECIALLY WORKING BACKWARDS AND SPATIAL VISUALIZATIONS) ~	NONE ~	NONE ~
20 MINUTES	<u>CALCULATOR PATHS</u> ~ CALCULATOR ASSISTED LOGICAL THINKING (p 235) ~ HELPS WITH ESTIMATION AND STRATEGY DEVELOPMENT ~	CALCULATOR PATHS INSTRUCTION SHEET ~	MARKERS ~ GAMEBOARDS ~
25 MINUTES	<u>USEFUL MATH SKILLS</u> ~ CAREER USES OF MATH (p. 271) DISCUSS RESULTS AFTERWARDS AND ANNOUNCE NEXT WEEK'S CAREER PANEL ~	USEFUL MATH SKILLS INSTRUCTIONS AND ANSWERS	USEFUL MATH SKILLS ANSWER SHEETS
10 MINUTES	ESTIMATE FOR TODAY ~ PICK SOMETHING FOR EVERYONE TO ESTIMATE ~ LIKE THE LENGTH OF CHALK TO THE NEAREST HALF CENTIMETER. RECORD ESTIMATES ~ HAVE CLASS COMPUTE MEAN ~ MEDIAN ~ AND MODE (p 141 IF THIS IS THE FIRST TIME FINDING THESE AVERAGES AS A CLASS) ~ MEASURE THE OBJECT ~ REPORT ~	NONE ~	OBJECT OR QUANTITY TO USE FOR ESTIMATION ~
15 MINUTES	<u>TAX COLLECTOR</u> ~ FACTORS ~ PRIME NUMBERS ~ ESTIMATION AND STRATEGY IN A GAME FOR 2 (p 67)	TAX COLLECTOR INSTRUCTIONS ~	SETS OF NUMBER SQUARES 1-24 (CUT FROM INDEX CARDS OR CALENDARS, GAMEBOARDS ETC) ~
5 MINUTES	SUMMARIZE SESSION ~		
	GIVE HOMEWORK ~ PUZZLE / CUT-A-CARD ~ THIS WEEK WAS MOSTLY NUMBERS AND LOGIC ~ NEXT WEEK GEOMETRY & CAREERS ~	CUT-A-CARD INSTRUCTIONS ~	CUT-A-CARD SAMPLE
	EVALUATION AND COMMENTS		PAPER ~

Mathematics Generally Covered
at Various Grade Levels
and
Resource List for
Parents and Teachers

MATHEMATICS GENERALLY COVERED

The following pages list the mathematics generally covered in the grade levels from Kindergarten through eighth grade.

Your own school or district will have its own list of skills for each grade level, which may look somewhat different. There is no hard and fast rule about the age at which students should learn most topics, and lists such as these should be considered general guidelines rather than absolute musts.

We have tried to emphasize the development of topics we feel are most important, and for that reason have placed application of mathematics in the context of daily living at the top of the list. Estimation is also mentioned repeatedly throughout, since it can give students such power in applying mathematics.

We also believe that children should be allowed to use calculators in the same way adults would—to take the drudgery out of long and tedious calculations. This requires that children be taught how to use calculators with the same seriousness that they have in the past been taught arithmetic skills. It also suggests, for example, that once the long division algorithm is understood, there is no benefit from doing page after page of long division problems. The time is better spent on using new word problem solution strategies, learning how to make another kind of graph, finding patterns in mathematics tables, or any of thousands of other fascinating activities.

It is extremely important that children be allowed to proceed at their own pace and not be forced to conform to this or any other list. If the list says, for example, that students should be able to count by twos, threes, fours, fives, and tens, but your child is really struggling with counting by ones, please do not press him or her to do more. Use lists with caution!

Mathematics Generally Covered in Kindergarten

Applications
☐ Talking about mathematics used in our daily lives

Numbers
☐ Learning to estimate how many
☐ Counting objects, up to about 15 or 20
☐ Putting out objects to match a number
☐ Comparing two sets of objects
☐ Recognizing numerals up to 20
☐ Writing numerals, 0 through 9
☐ Learning about ordinal numbers, such as first, second, third

Measurement
☐ Estimating and comparing:

Taller or shorter

Longer or shorter

Largest or smallest

Heavier or lighter

Geometry
☐ Recognizing and classifying colors and simple shapes

Patterns
☐ Recognizing simple patterns, continuing them, and making up new patterns

Probability and Statistics
☐ Making and talking about simple graphs of everyday things, such as birthdays, pets, food, and so on

Mathematics Generally Covered in First Grade

Applications

☐ Talking about mathematics used in daily living

☐ Learning strategies such as using manipulatives or drawing diagrams to solve problems

Arithmetic and Numbers

☐ Practicing estimation skills

☐ Counting through about 100

☐ Recognizing, writing, and being able to order numbers through about 100

☐ Counting by twos, fives and tens

☐ Using ordinal numbers, such as first, second, tenth, and so on

☐ Learning basic addition and subtraction facts up to $9+9=18$ and $18-9=9$

☐ Developing understanding of place value using tens and ones with manipulatives, including base ten blocks, Cuisenaire rods, abaci, play money, and so on

☐ Developing the concept of fractional values such as halves, thirds, and fourths

Measurement

☐ Telling time to the hour or half-hour (but don't press if not mastered)

☐ Recognizing and using calendars, days of the week, months

☐ Estimating lengths and measuring things with non-standard units, such as how many handprints across the table

☐ Understanding uses and relative values of pennies, nickels, dimes

Geometry and Patterns

☐ Working with shapes, such as triangles, circles, squares, rectangles

☐ Recognizing, repeating, and making up geometric and numeric patterns

Probability and Statistics

☐ Making and interpreting simple graphs, using blocks or people, of everyday things, such as color preferences, number of brothers and sisters, and so on

Mathematics Generally Covered in Second Grade

Applications
- ☐ Talking about mathematics used in daily life
- ☐ Creating and solving word problems in measurement, geometry, probability, and statistics, as well as arithmetic
- ☐ Practicing strategies for solving problems, such as drawing diagrams, organized guessing, putting problems into own words, and so on

Numbers
- ☐ Practicing estimation skills
- ☐ Reading and writing numbers through about 1,000, playing around with up to 10,000
- ☐ Counting by twos, fives, and tens, and maybe some other numbers for fun
- ☐ Learning about odd and even numbers
- ☐ Using ordinal numbers such as first, second, tenth
- ☐ Identifying fractions such as halves, thirds, quarters
- ☐ Understanding and using the signs for *greater than* (>) and *less than* (<)

Arithmetic
- ☐ Knowing addition and subtraction facts through $9+9=18$ and $18-9=9$
- ☐ Estimating answers to other addition and subtraction problems
- ☐ Practicing addition and subtraction with and without regrouping (carrying), such as:

$$\begin{array}{r} 2\ 7 \\ +\ 2 \\ \hline 2\ 9 \end{array} \qquad \begin{array}{r} 2\ 7 \\ +\ 8 \\ \hline 3\ 5 \end{array} \qquad \begin{array}{r} 2\ 7 \\ -\ 2 \\ \hline 2\ 5 \end{array} \qquad \begin{array}{r} 2\ 7 \\ -\ 8 \\ \hline 1\ 9 \end{array}$$

- ☐ Adding columns of numbers, such as

$$\begin{array}{r} 2 \\ 8 \\ 9 \\ +\ 7 \\ \hline \end{array}$$

- Exploring uses of a calculator
- Being introduced to multiplication and division

Geometry

- Finding congruent shapes (same size and shape)
- Recognizing and naming squares, rectangles, circles, and maybe some other polygons
- Informally recognizing lines of symmetry
- Reading and drawing very simple maps

Measurement

- Practicing estimation of measurements—how many toothpicks long is the table?
- Comparing lengths, areas, weights
- Measuring with non-standard units, beginning to use some standard units such as inches or centimeters
- Telling time to the nearest quarter-hour, maybe to the minute
- Making change with coins and bills, doing money problems with manipulatives
- Knowing days of the week and months, and using the calendar to find dates

Probability and Statistics

- Making and interpreting simple graphs, using physical objects or manipulatives
- Doing simple probability activities

Patterns

- Working with patterns of numbers, shapes, colors, sounds, and so on, including adding to existing patterns, completing missing sections, making up new patterns

Mathematics Generally Covered in Third Grade

Applications

- ☐ Talking about mathematics seen in students' lives
- ☐ Creating, analyzing, and solving word problems in all of the concept areas
- ☐ Practicing a variety of problem-solving strategies with problems of more than one step

Numbers

- ☐ Practicing estimation skills with all problems
- ☐ Reading and writing numbers through about 10,000, exploring those beyond 10,000
- ☐ Counting by twos, threes, fours, fives and tens and other numbers
- ☐ Naming and comparing fractions such as 1/2 is greater than 1/4
- ☐ Identifying fractions of a whole number such as 1/2 of 12 is 6
- ☐ Exploring concepts of decimal numbers such as tenths and hundredths, using money to represent values
- ☐ Using the signs for greater than (>) and less than (<)

Arithmetic

- ☐ Learning how to use calculators effectively
- ☐ Using calculators to solve problems
- ☐ Continuing to practice basic addition and subtraction facts and simple addition and subtraction problems
- ☐ Doing larger and more complicated addition and subtraction problems

$$
\begin{array}{r} 3\ 8\ 9\ 7 \\ +\ 8\ 3\ 4\ 2 \end{array}
\qquad
\begin{array}{r} 8\ 3\ 4\ 2 \\ -\ 3\ 8\ 9\ 7 \end{array}
$$

- ☐ Beginning to learn multiplication and division facts through $9 \times 9 = 81$ and $81 \div 9 = 9$
- ☐ Beginning to learn multiplication and division of two- and three-digit numbers by a single-digit number

$$
\begin{array}{r} 2\ 7 \\ \times\ 3 \end{array}
\qquad
\begin{array}{r} 1\ 2\ 4 \\ \times\ 8 \end{array}
\qquad
6\overline{)2\ 4}\,^{4}
$$

- ☐ Learning about remainders

$$
\begin{array}{r}
4\ \text{R}\ 1 \\
7\overline{)2\ 9} \\
2\ 8 \\
\hline
1
\end{array}
$$

Geometry

- ☐ Recognizing and naming shapes such as squares, rectangles, trapezoids, triangles, circles, and three-dimensional objects such as cubes, cylinders, and the like
- ☐ Identifying congruent shapes (same size and shape)
- ☐ Recognizing lines of symmetry, and reflections (mirror images) and translations (movements to a different position) of figures
- ☐ Reading and drawing simple maps, using coordinates
- ☐ Learning about parallel (||) and perpendicular (⊥) lines

Measurement

- ☐ Estimating before doing measurements
- ☐ Using non-standard and some standard units to measure:
  - ☐ Length
    (toothpicks, straws, paper strips, string lengths, and so on)
    (centimeters, decimeters, meters, inches, feet, yards)
  - ☐ Perimeters
    (same as length)
  - ☐ Area
    (square units)
    (paper squares, tiles, and so on)
    (square centimeters, meters, inches, feet, yards)
  - ☐ Weight
    (paper clips, rocks, blocks, beans, and so on)
    (grams, kilograms, ounces, pounds)
  - ☐ Volume and capacity
    (blocks, rice, beans, water; in cans, paper cups, and so on)
    (liters, cubic centimeters, cups, gallons, pints, quarts)
  - ☐ Temperature
    (° Celsius, ° Fahrenheit)
- ☐ Telling time, probably to the nearest minute
- ☐ Continuing to use money to develop understanding of decimals
- ☐ Using calendars

Probability and Statistics

- ☐ Being introduced to probability concepts, such as the chance of something happening
- ☐ Using tally marks, collecting and organizing informal data
- ☐ Making, reading, and interpreting simple graphs

Patterns

- ☐ Continuing to work with patterns, including those found on addition and multiplication charts

Mathematics Generally Covered in Fourth Grade

Applications

- ☐ Talking about uses of mathematics in students' lives and in their futures
- ☐ Creating, analyzing, and solving word problems in all of the concept areas
- ☐ Using a variety of problem solving strategies to solve problems with multiple steps
- ☐ Working in groups to solve complex problems
- ☐ Using calculators for problem solving
- ☐ Developing formal and informal mathematical vocabulary

Numbers and Operations

- ☐ Practicing rounding and estimation skills with all problems
- ☐ Using calculators with some proficiency for all operations
- ☐ Reading and writing numbers to 10,000 and beyond
- ☐ Learning about special numbers, such as primes, factors, multiples, square numbers
- ☐ Recognizing equivalent fractions, such as $1/2 = 2/4$
- ☐ Finding fractions of whole numbers, such as $1/8$ of $72 = 9$
- ☐ Maintaining and extending work with the operations of addition, subtraction, multiplication, and division of whole numbers
- ☐ Adding and subtracting simple decimal numbers
- ☐ Learning about simple percents such as 10%, 50%, and 100%

Geometry

- ☐ Using geometric shapes to find patterns of corners, diagonals, edges, and so on
- ☐ Recognizing right angles, exploring terminology of other angles
- ☐ Continuing to explore ideas connected with symmetry
- ☐ Reading and drawing simple maps, using coordinates
- ☐ Exploring terminology and uses of coordinate grid
- ☐ Identifying parallel (||) and perpendicular (⊥) lines
- ☐ Exploring how different shapes fill a flat surface (tiling)

Measurement

- ☐ Estimating before measuring
- ☐ Using non-standard and standard units to measure length, area, volume, weight, temperature
- ☐ Telling time for a purpose

- [ ] Making simple scale drawings
- [ ] Exploring terminology and uses of geometric grid

Probability and Statistics

- [ ] Using sampling techniques to collect information or conduct a survey
- [ ] Discussing uses and meanings of statistics, such as how to make a survey fair, how to show the most information, how to find averages
- [ ] Making, reading, and interpreting graphs
- [ ] Performing simple probability experiments, discussing results

Mathematics Generally Covered in Fifth and Sixth Grades

Applications

- ☐ Talking about mathematics used in present and future lives
- ☐ Creating, analyzing, and solving word problems in all of the concept areas
- ☐ Using a variety of problem solving strategies to solve problems with multiple steps
- ☐ Working in groups to solve complex problems
- ☐ Using calculators for problem solving
- ☐ Developing mathematical vocabulary

Numbers and Arithmetic

- ☐ Practicing rounding and estimation skills with all problems
- ☐ Using calculators effectively for appropriate problems
- ☐ Expanding understanding and use of special numbers such as primes, composite numbers, square and cubic numbers, common divisors, common multiples
- ☐ Increasing understanding of fraction relationships:

  Comparisons, such as $2/3 > 1/2$

  Equivalence, such as $2/3 = 4/6$

  Reducing, such as $10/20 = 1/2$

  Relating mixed numbers and improper fractions, such as $2\ 1/3 = 7/3$

- ☐ Developing skills in adding, subtracting, multiplying, dividing fractions (mastery not expected)
- ☐ Maintaining skills in basic addition, subtraction, multiplication, division of whole numbers
- ☐ Adding, subtracting, multiplying, and dividing decimal numbers
- ☐ Computing percents, and relating percents to fractions and decimals
- ☐ Developing some understanding of ratio and proportion
- ☐ Exploring scientific notation such as $3 \times 10^8 = 300,000,000$

Geometry

- ☐ Using the concept of parallel ($||$) and perpendicular ($\perp$) lines
- ☐ Measuring and drawing angles of various kinds
- ☐ Understanding circle relationships, including diameter, circumference, radius
- ☐ Recognizing shapes that are congruent, or the same size and shape

- Recognizing shapes that are similar, or the same shape but a different size
- Developing understanding of symmetry, reflections, and translations of figures
- Drawing constructions such as equal line segments or perpendicular bisectors
- Understanding coordinate graphing
- Drawing and reading maps
- Doing perspective drawing

Measurement

- Continuing to use hands-on tools of measurement, estimating first in all cases, for:

  Length

  Area

  Volume and capacity

  Mass or weight

  Temperature—Celsius and Fahrenheit

  Telling time accurately

Probability and Statistics

- Performing and reporting on a variety of probability experiments
- Collecting and organizing data
- Displaying data in graphic form, such as bar, picture, circle, line, and other graphs
- Beginning to develop understanding of statistical ideas such as mean, median, and mode

Mathematics Generally Covered in Seventh and Eighth Grades

Applications

- ☐ Talking about uses of mathematics and its importance to students' present and future lives (especially important for female and minority students)
- ☐ Creating, analyzing, and solving word problems in all of the concept areas
- ☐ Using a variety of problem solving strategies to solve problems with multiple steps
- ☐ Working in groups to solve complex problems
- ☐ Using calculators for problem solving
- ☐ Developing mathematical vocabulary

Numbers and Arithmetic

- ☐ Practicing rounding and estimation skills with all problems
- ☐ Using calculators with proficiency for appropriate problems
- ☐ Expanding understanding and use of special numbers such as primes, composite numbers, square and cubic numbers, common divisors, common multiples
- ☐ Increasing understanding of fraction relationsips:

    Comparisons, such as $2/3 > 1/2$

    Equivalence, such as $2/3 = 4/6$

    Reducing, such as $10/20 = 1/2$

    Relating mixed numbers and improper fractions, such as $2\ 1/3 = 7/3$
- ☐ Adding, subtracting, multiplying, and dividing fractions
- ☐ Adding, subtracting, multiplying, and dividing decimal numbers
- ☐ Maintaining skills in basic addition, subtraction, multiplication, division of whole numbers
- ☐ Computing percents, and relating percents to fractions and decimals
- ☐ Understanding of ratio and proportion
- ☐ Using scientific notation such as $3 \times 10^8 = 300,000,000$
- ☐ Learning about positive and negative numbers
- ☐ Finding greatest common factors (GCF) and least common multiples (LCM)
- ☐ Finding square roots
- ☐ Learning about special number relationships such as 2/5 is a reciprocal of 5/2

Geometry

- ☐ Using the concept of parallel (||) and perpendicular (⊥) lines
- ☐ Measuring and drawing angles of various kinds
- ☐ Understanding circle relationships, including diameter, circumference, radius
- ☐ Using correct formulas to calculate areas of rectangles, triangles, circles, and so on
- ☐ Recognizing shapes that are congruent, or the same size and shape
- ☐ Recognizing shapes that are similar or the same shape but a different size
- ☐ Developing understanding of symmetry, and reflections, and translations of figures
- ☐ Making constructions such as equal line segments or perpendicular bisectors
- ☐ Understanding coordinate graphing
- ☐ Drawing and reading maps
- ☐ Doing scale and perspective drawing

Measurement

- ☐ Continuing to have hands-on experiences with the tools of measurement, estimating first in all cases, for:

  Length

  Area

  Volume and capacity

  Mass or weight

  Temperature—Celsius and Fahrenheit

  Telling time accurately

Probability and Statistics

- ☐ Performing and reporting on a variety of probability experiments
- ☐ Collecting and organizing data
- ☐ Displaying data in graphic form, such as bar, picture, circle, line, and other graphs
- ☐ Developing understanding of statistical ideas such as mean, median, and mode

RESOURCE LIST FOR
PARENTS AND TEACHERS

CODE:
P (primary—K–3); E (elementary—4–6);
M (middle/jr. high—6–9)
*Upper elementary and middle school students will also enjoy reading these books.

EM   Afflack, Ruth. *Beyond Equals*. Oakland, CA: The Math/Science Resource Center, 1982.

PEM   Alper, Lynne, and Holmberg, Meg. *Parents, Kids, and Computers*. Berkeley, CA: Sybex, Inc. 1984.

P   Baratta-Lorton, Mary. *Workjobs... For Parents*. Menlo Park, CA: Addison-Wesley Publishing Co., 1972.

PE   Baratta-Lorton, Mary. *Mathematics Their Way*. Menlo Park, CA: Addison-Wesley Publishing Co., 1976.

EM   Bezuszka, Stanley; Kenney, Margaret; and Silvey, Linda. *Designs for Mathematical Patterns*. Palo Alto, CA: Creative Publications, 1978.

EM*   Burns, Marilyn. *Math For Smarty Pants*. Boston, MA: Little, Brown, and Company, 1982.

EM*   Burns, Marilyn. *The Book of Think*. Boston, MA: Little, Brown, and Company, 1976.

EM   Burns, Marilyn. *The Good Time Math Event Book*. Palo Alto, CA: Creative Publications, Inc., 1977.

EM*   Burns, Marilyn. *The I Hate Mathematics Book*. Boston, MA: Little, Brown, and Company, 1975.

EM*   Burns, Marilyn; Weston, Martha; and Allison, Linda. *Good Times: Every Kid's Book of Things to Do*. New York: Bantam Books, 1979.

EM   Cook, Marcy. *Mathematics Problems of the Day*. Palo Alto, CA: Creative Publications, Inc., 1982.

PEM   Downie, Diane; Slesnick, Twila; and Stenmark, Jean K. *Math for Girls and other Problem Solvers.* Berkeley, CA: Lawrence Hall of Science, University of California, 1981.

M   Fisher, Lyle. *Super Problems.* Palo Alto, CA: Dale Seymour Publications, 1982.

EM   Fraser, Sherry, Project Director. *SPACES: Solving Problems of Access to Careers in Engineering and Science.* Berkeley, CA: Lawrence Hall of Science, University of California, 1982.

PEM   Harnadek, Anita. *Mindbenders Levels A,B,C.* Pacific Grove, CA: Midwest Publications, 1978.

EM   Kaseberg, Alice; Kreinberg, Nancy; and Downie, Diane. *Use EQUALS to Promote the Participation of Women in Mathematics.* Berkeley, CA: Lawrence Hall of Science, University of California, 1980.

P   Meiring, Stephen P. *Parents and the Teaching of Mathematics.* Columbus, Ohio: Ohio Department of Education, 1980.

EM   Meyer, Carol, and Sallee, Tom. *Make it Simpler: A Practical Guide to Problem Solving in Mathematics.* Menlo Park, CA: Addison-Wesley Publishing Company, 1983.

EM   Miller, Con. *Calculator Explorations and Problems.* New Rochelle, New York: Cuisenaire Company of America, Inc., 1979.

EM   Pedersen, Jean J., and Armbruster, Franz O. *A New Twist: Developing Arithmetic Skills Through Problem Solving.* Menlo Park, CA: Addison-Wesley Publishing Company, 1979.

EM   Rand, Ken. *Point-Counterpoint: Graphing Ordered Pairs.* Palo Alto, CA: Creative Publications, Inc., 1979.

M   Saunders, Hal. *When Are We Ever Gonna Have to Use This?* Palo Alto, CA: Dale Seymour Publications, 1981.

P   Schreiner, Bryson. *Arithmetic Games and Aids for Early Childhood.* Hayward, CA: Activity Resources Company, Inc., 1974.

EM   Seymour, Dale. *Developing Skills in Estimation Book A.*
          Palo Alto, CA: Dale Seymour Publications, 1981.

M   Seymour, Dale. *Developing Skills in Estimation Book B.*
          Palo Alto, CA: Dale Seymour Publications, 1981.

M   Seymour, Dale. *Favorite Problems.* Palo Alto, CA: Dale
          Seymour Publications, 1984.

EM   Seymour, Dale. *Problem Parade.* Palo Alto, CA: Dale
          Seymour Publications, 1984.

EM   Seymour, Dale. *Visual Thinking Cards.* Palo Alto, CA:
          Dale Seymour Publications, 1983.

P   Sharp, Evelyn. *Thinking is Child's Play.* New York: Avon
          Books, 1969.

EM   Shulte, A.P., and Choate, S. *What Are My Chances?*
          Book A. Palo Alto, CA: Creative Publications, Inc.,
          1977.

M   Shulte, A.P., and Choate, S. *What Are My Chances?*
          Book B. Palo Alto, CA: Creative Publications, Inc.,
          1977.

PEM   Skolnick, Joan; Langbort, Carol; and Day, Lucille.
          How to Encourage Girls in Math and Science—Strategies
          for Parents and Educators. Englewood Cliffs, New
          Jersey: Prentice-Hall, Inc., 1982.

P   Sprung, Barbara; Campbell, Patricia B.; and Froschl,
          Merle. *What Will Happen if . . . Young Children and the*
          Scientific Method. New York, NY: Education Equity
          Concepts, Inc., 1985.

EM   Stonerod, Dave. *Friendly Games to Make and Learn.*
          Hayward, CA: Activity Resources Company, Inc.,
          1975.

EM   Thiagarijan, Sivasailam, and Stolvitch, Harold. *Games*
          with the Pocket Calculator. Menlo Park, CA: Dymax,
          1976.

PE   Wirtz, Robert. *Making Friends with Numbers, Kits I and II.*
          Monterey, CA: Curriculum Development Associates,
          1977.

P   Zaslavsky, Claudia. *Preparing Young Children for Math:*
          A Book of Games. New York: Schocken Books, 1979.

ADDRESSES OF PUBLISHERS

ACTIVITY RESOURCES CO., INC.
P.O. Box 4875
Hayward, CA 94540

ADDISON-WESLEY PUBLISHING CO.
2725 Sand Hill Road
Menlo Park, CA 94025

AVON BOOKS
1790 Broadway
NewYork, N.Y. 10019

BANTAM BOOKS
666 Fifth Ave.
New York, N.Y. 10019

CREATIVE PUBLICATIONS
P.O. Box 10328
Palo Alto, CA 94303

CUISENAIRE COMPANY OF AMERICA
12 Church St., Box D
New Rochelle, N.Y. 10805

CURRICULUM DEVELOPMENT
 ASSOCIATES
787 Foam St.
Monterey, CA 93940

DALE SEYMOUR PUBLICATIONS
P.O. Box 10888
Palo Alto, CA 94303

EDUCATION EQUITY CONCEPTS, INC.
440 Park Ave. S.
New York, NY 10016

LAWRENCE HALL OF SCIENCE
University of California Berkeley
Berkeley, CA 94720

DYMAX
Box 310
Menlo Park, CA 94025

LITTLE, BROWN, AND COMPANY
34 Beacon St.
Boston, MA 02106

MATH/SCIENCE RESOURCE CENTER
Mills College
Oakland, CA 94613

MIDWEST PUBLICATIONS
P.O. Box 448
Pacific Grove, CA 93950

NATIONAL COUNCIL OF TEACHERS
 OF MATHEMATICS
1906 Association Drive
Reston, Virginia 22091

OHIO DEPARTMENT OF EDUCATION
65 South Front St.
Columbus, Ohio 43215

PRENTICE-HALL, INC.
Route 9W
Englewood Cliffs, N.J. 07632

SCHOCKEN BOOKS, INC.
200 Madison Ave.
New York, N.Y. 10016

SYBEX INC.
2344 Sixth St.
Berkeley, CA 94710

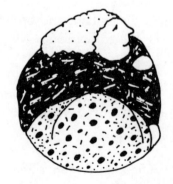

INDEX